STRUCTURED SYSTEM ANALYSIS:
A New Technique

STRUCTURED SYSTEM ANALYSIS:
A New Technique

Barbara F. Medina

GORDON AND BREACH SCIENCE PUBLISHERS
New York   London   Paris

Copyright © 1981 by Gordon and Breach, Science Publishers, Inc.

Gordon and Breach, Science Publishers, Inc.
One Park Avenue
New York, N.Y. 10016

Gordon and Breach Science Publishers Ltd.
42 William 1V Street.
London, WC2N 4DE

Gordon and Breach
7-9 rue Emile DuBois
Paris 75014

Library of Congress Cataloging in Publication Data

Medina, Barbara F. 1931—
    Structured system analysis.

    Bibliography: p.
    Includes index.
    1. System analysis. I. Title.
QA402.M43   003   80-8819
ISBN 0-677-05570-6

Library of Congress catalog card number 80-8819. ISBN 0 677 05570 6. All rights reserved. No part of this book may be reproduced or utilized in any form or by any means, electronic or mechanical, including photocopying, recording, or by any information storage or retrieval system, without permission in writing from the publishers. Printed in the United States of America.

ACKNOWLEDGEMENTS

I want to take this opportunity to thank three of my colleagues for their assistance in helping me with this book. Steven Schroeder and David Straub for their reviewing the text and adding some valuable insights to my thinking. Gary Slater for drawing the figures and providing editorial assistance. No acknowledgement would be complete without the mention of Anne Mestan whose fine secretarial skills and patience in typing the original manuscript provided such fine support for this project.

PREFACE

Much of what is stated in this section might have been included in Chapter I of the book. Since the objective of this section is to acquaint you with my reasoning prior to writing the book, to tell you what you will find in the book, and to tell you what audience I feel would benefit most from the concepts I'm trying to prove, it seemed more consistent with my theories about systems to place these statements outside the main body of the book.

Let's start with the term system. When I first considered trying to prove that certain types of things we call systems could be examined much as we explore geometric properties of lines and triangles, I thought about using another word rather than system. The term system means many things to different people. Training, background, line of work, all influence what the word means to you. Just examining the definition of the noun, system, in an unabridged dictionary illustrates this point. For example, the definition for system from the 1966 edition of The Random House Dictionary gives 14 definitions for the noun system. The disciplines cited for special definitions are astronomy, biology, geology, physical chemistry. Since this is a 1966 dictionary the adoption of this word and the emphasis on it by people who call themselves system analysts is not even mentioned. Thus, no special definition for computer science or more narrowly automated information processing is offered.

What convinced me to use the word, in spite of the impreciseness, is that it does connote: internal interactions, something that is to be considered as one unit, and an entity with a potential for ordering in everyone's mind. This is the type of entity I want to examine and have the reader think about.

For those of us who have been involved in automated information processing, the confusion of meaning goes way beyond what would appear in a standard dictionary. We talk of a computer as a system, the program that coordinates the components of the computer as a system, the instructions we write to process information as a system, and the model that we use to simulate a real life situation as a system. When I looked at all these 'systems' it became apparent that, if their boundaries were defined correctly, all are systems as I want to use the term. Let me elaborate on this a bit. The computer, if you mean all the hardware and software that make up a piece of equipment that can accept, process, and report information, can be a system as I want to define system. Of course there are computers that are cumbersomely organized and, hence, inefficient and difficult to use. This does not change the fact that they are systems. The operating 'system' of the computer system is itself a system, and a subsystem of the larger system, "the computer." A simulation is simply an approximation and/or simplification of the real work system, and if all the critical parts of the real world system are incorporated in the model, there is no problem in using a simulation to evaluate a system. The problems usually occur in determining what is critical. The power of simulation as a tool is that if you omit a critical subsystem or factor it becomes apparent when you compare it against the system you are trying to simulate. Unfortunately, individuals who use this tool often forget that they are looking at an approximation and will try and make reality reflect the output of their model instead of vice versa.

A perfect example of this has occurred in Pennsylvania where the models designed by the National Consortium for Higher Education Management Systems (NCHEMS) for planning in higher education are being used for allocation of funds. All the precautionary statements of the developers have been ignored and colleges are to be funded on the basis of "an equity line" that is a curve fitting approximation of information that comes from an approximate model of a system whose interactions are not well understood.

Thus, I have concluded that although there is no agreement on the exact meaning of the term, and it is abused and often misused, it is a useful term. There is a good deal of precedent in all disciplines to take a widely used term and precisely define it for the discipline. A great many things we want to study today cross classical discipline lines. This is certainly true for most things we call systems. Hence, I am proposing a definition for system that could be useful to many disciplines that have a need to study entities whose information from without can be obtained once at the beginning of a segment of time, and under 'steady state' (i.e., usual conditions) can complete the cycle without transmitting information externally. Within these limitations there are many things we want to study that can be considered a system, for example: organizations, computers, ecological systems, transportation systems, etc.

How we look at this system will have to change if we want to include within the system's boundaries all the components that affect them and cause interactions. For example, in analyzing an organization as a system, all the components that would have to be included within the boundaries of the organization's universe would have to be included in our analysis. This would be their suppliers, employees, procedures, organization structure, and their customers. There are external factors that affect the organization, but these can be excluded from the system as they meet the criteria that information from them can be entered once at the beginning of a cycle, and under 'steady state' condition they will not change during the length of a cycle. For example, many organizations have to deal with governmental regulations in their day-to-day operations. However, although these regulations certainly do change, and they directly affect the organization's way of handling their affairs under steady state conditions, the organization can usually go through one cycle, let's arbitrarily choose fiscal year, without change.

In order to try and change the way one looks at systems, I have stated my premises about systems in Chapter I. Starting with these premises in Chapter I, the book is then divided into four additional chapters. Chapter II defines a system and subsystem and develops two theorems about their relationship to each other.

Chapter III defines modules and develops the relationships between modules and subsystems. It develops theorems that set guidelines for the internal structure of a system. It introduces the concept of information and its attributes.

The interchange of information within a system is then

explored in Chapter IV. These concepts of information transfer are used to show how level within a system is determined.

Finally in Chapters V and VI, illustrative applications of the theorems and guidelines developed in the preceding chapters are given. Because of the pivotal nature of the theorems developed in Chapter IV, the problems associated with bad system analysis are described in the previous chapters and the solutions have to wait until these chapters.

Much has been written about general systems theory in the last thirty years. A great deal of my own thinking on this subject has been aided by some of the classical work in this field. My own interest has been narrower than is reflected in most of the writing in general systems theory. I have been more interested in providing guidelines for the decomposition of systems into useful subunits than in defining classes of systems that might be useful for a generalized approach to science. However, I feel I have been influenced by these writers' thinking and insights and must acknowledge my indebtedness to the authors of works in general systems theory.

References to articles and books related to this topic will be found in the bibliography, but I want to acknowledge a few very important names here. Since von Bertanffy articles on general system theory first appeared in 1951, several authors have attempted to apply his theories to specific disciplines: Boulding, Johnson, Kast and Rosenzweig to management and organization problems; Laszlo to philosophy. Weizenbaum has written a very thoughtful book on the role of automation.

Reading these articles and books over a period of years has forced me to be more specific about my own line of reasoning. I wanted to be able to provide guidelines for people trying to analyze a system. I have written articles where I have presented guidelines, but these guidelines were based only on my own experience in successfully analyzing a system.

At a working meeting in Hungary, in January of 1979, on the Social Impact of Technology, it became apparent to all of us who attended (from about 50 countries) that we had to define our terms before we could communicate. The term system was one that was used frequently and we became aware it had different implications for each speaker. This was particularly true because the group was made up of both information and process specialists. Once we had agreed upon terminology, the discussions became more fruitful.

Upon returning home I found myself rethinking an idea I had about five years ago. Why not try and define terms and then build the guidelines in a logical fashion by proving theorems based on definitions and on the proof of previous theorems. I found I was able to take this approach only if I restricted the term system to an entity which, during an identifiable period of time, received no information from outside the system. Since most systems of this type that are of interest contain what von Bertanffy defined as the class of open system, and Laszlo used man as an example of an open system (a system which continuously receives information from an external source), I tried to take this approach also.

I found it confusing to try to use this terminology in the theorems I developed, so I have substituted the term module for the type of entity which receives continuous information from outside. I chose module for several reasons: I found it essential to differentiate between entities that simply continuously process information and those that coordinate interactions within the system. It is also important to have one entity that the property of being closed or open is not the determining factor in defining the entity.

Hence I have reserved the term system for entities that have a definable cycle time and the term module for entities that either process information or coordinate interactions within a system. Thus there is one type of system being discussed and two types of modules. The module that coordinates information can also have the properties of a system but a system by my definitions cannot simply process information: it must coordinate interactions.

This leads us to who can use the techniques developed in this book. My own feeling is anyone who wants to analytically study a system. The insights one would gain can then be used to help resolve problems or answer questions about the system. However, I want to make it very clear that one would simply gain insights, and that I am not recommending that the theorems developed in this book can be used as one more tool to supersede human intelligence.

Almost any system we want to study contains humans. People are amazingly diverse in the reactions to any situation. They, of course, will affect the interactions in any system. No one can lay down a fixed set of rules as to how they will behave in a particular situation. Thus, if you use a tool such as statistical analysis, simulation, or system analysis, as if it provides you with biblical truths, you are deceiving yourself and the people you have convinced with your 'bottom line' argument. Hopefully, enough people will be trained to be able to analyze a situation so that it is not so easy for the experts to deceive them with their mathematical tricks. To help obtain this knowledgeable public was my motivation in developing the theory found in this book.

Barbara Medina
Director of Information Services
University of Maryland Baltimore County

CONTENTS

PREFACE vii

I   SYSTEMS, SUBSYSTEMS, MODULES 1

II   WHAT IS A SYSTEM? A SUBSYSTEM? 6
       Theorem 1   Cycle Times 9
       Theorem 2   Hierarchical order of subsystems 10

III   INTERNAL STRUCTURE OF SYSTEMS 16
       Theorem 3   Systems with one module 17
       Theorem 4   Subsystem can be coordinator module .. 18
       Theorem 5   INFO value calculated in one module ... 19
       Theorem 6   One first-level coordinator module from a system with subsystems 20
       Theorem 7   A system with one coordinator module has no subsystems 21
       Theorem 8   System with multiple coordinator modules can be decomposed into subsystems 22
       Theorem 9   System with n coordinator modules has no more than $n-1$ subsystems . 22
       Theorem 10   Modules are systems only if they have cycle times 23
       Theorem 11   Systems with one coordinator module are hierarchical 23
       Theorem 12   Systems with one coordinator, process modules and subsystems are hierarchical 24

IV   INFORMATION TRANSFER AND LEVELS 36
       Theorem 13   Information determines hierarchical placement 37

	Theorem 14	Values of input switches or flags must be set hierarchically above where they are used 38
	Theorem 15	The value of global INFOs must be established in systems with longer cycle times than the user system 39
	Theorem 16	The number of input records is determined by global item placement 40

V   SOLUTIONS: AN ORGANIZATION AS A SYSTEM .. 47

VI   SOLUTIONS: MODELING A COMPLEX SYSTEM ... 63

BIBLIOGRAPHY   77

I
SYSTEMS, SUBSYSTEMS, MODULES

Anyone who has attended a seminar on systems, or read a book about systems design, probably has come to the realization that the term system is loosely used and means different things not only to the general public, but to system analysts. Most system analysts would agree on certain things. There are entities that can be analyzed and handled as one system, and that if you can decompose these entities into smaller independent units which are referred to somewhat arbitrarily as subsystems or modules, it is easier to work with them. Finally, at least as far as designing automated systems is concerned, a hierarchical system design is desirable.

There seems to be no agreement on several very basic ideas about systems in spite of all that has been written and discussed about them. Exactly when should we use the term system to describe an entity? Can all systems be broken up into smaller parts and, if so, how should they be decomposed? Are the words system and module interchangeable? If not, what is the difference between the two concepts? How does one set guidelines for good system design? Why are so many systems unresponsive to their users, difficult to test or validate, unsafe, etc.?

Many attempts have been made to precisely define the term system. In spite of this, the word is used to signify almost anything. Most good system designs seem to occur because the designer has some mystic ability to design good systems.

In order to at least start to formalize the entire subject, it would be desirable to be able to develop guidelines and theorems about at least some types of well defined systems. This is what this book does.

It starts with the premise that closed, bounded systems are of interest and can be defined, and that all such systems possess the dimension time. Thus, if you are looking at a process or a function that either has no cycle time, or depends on continuous information from an outside source, or must provide continuous information to an outside source, you are not looking at a system. The phrase *cycle time* is defined as an identifiable period of time where all interactions within the system occur without a need for new information obtained from outside the system.

I have placed all the definitions of the terms used to prove the theorems in the book here at the beginning. They appear in the order that they are used in the book. Thus, you will find cycle time as the first definition. In addition, when they appear in the text they are italicized and identified as a definition of a term that will be referred to in the proofs. Before proceeding to the next premise, let's look at the definitions of the terms.

Cycle Time: An identifiable period of time where all interactions within the system occur without a new need for new information obtained from outside the system.

Entity: Anything that can be evaluated as a whole.

A System: The total entity to be evaluated. This must include all the people and things that interact within the system. All forms, reports and filing associated with the system (whether automated or not); and, in the case of automated systems, the programs. It must contain a structure that allows for one controlled entry to the system, and one exit from the system, and for instant interactions. The system must be capable of operating through one complete cycle on the entry information.

A *proper subsystem* of a system must itself be a system.

Hierarchy: A group of systems or modules organized into orders or ranks, each order or rank subordinate to the one above it.

Level: The hierarchical placement of a module or subsystem.

Module: Is an identifiable entity that can occur within a system. It can be an entity that coordinates the actions of other modules, or subsystems within the system, or it can accomplish a process. It may or may not possess a cycle time. Hence, there can be continuous input and/or output with other units of the system. Thus, there are two types of modules; each is defined below.

1. *Coordinator Module:* Is a module within a system which has the following attributes.
    a. Can evoke subsystems *or* modules within the system.
    b. Can pass information directly from one subsystem or module to another within the system, or place INFO values (see definition below) in a permanent record of the system.
    c. Make decisions based on flags or switches (see definition below) set by other modules or subsystems within the system or

by entry information.

2. *Process Module*: Is the module within a system which has the following attributes:
   a. Calculates an INFO (see definition below) value.
   b. Performs an iterative process.
   c. Places a value in an INFO that will be used as a flag or a switch.

Information (INFO): An INFO is a unique member of a system that has specific attributes for this system. The INFO's attributes can be defined by:

1. A constant
2. A decision table
3. An attribute table.
4. A logical relationship.
5. A relational relationship.
6. A probabilistic relationship.

INFO Attribute: Is the characteristic that uniquely defines the INFOs of a system.

Switch or a Flag: An INFO that will be used by the coordinator to determine the flow of a system.

Branch: A unique path from the top of a system to the end of a two-way path which ends in a module of the system.

Steady State: The condition that describes the interactions within a system uninterrupted by new entry INFO during a cycle of the system.

Global INFO: Is an INFO whose constant attributes will be used as entry information in more than one module of the system.

File: Is a related collection of global INFOs and their constant attributes.

The next premise is that a proper subsystem must itself be a system. That a subsystem and a module of a system can be two very different things. Both are useful to decompose a system into a workable and useful structure, but they are not necessarily the same thing. The distinction is made on the basis of the dimension time. A module of a system does not have to possess a cycle time; it can receive continuous input from the system and can produce continuous output. It need not be a system.

The final and possibly the most important premise is that systems, once they are defined properly, usually can be decomposed into smaller self-contained entities. There are, of course, entities whose internal interactions are so simple that there is no need or desire to break them up into smaller entities to study them. One example might be the automated system that finds the roots of a quadratic equation. Another might be a very small business organization. However, most of the things we want to study using system analysis techniques tend to be complex interacting entities that can only be analyzed if they are broken up into more manageable subdivisions which can then be reunited to reform the system.

The first chapter is meant to introduce the objective of the book and to defend the usefulness of even this limited approach to a uniform system theory. Limited, because all we will discuss is closed and bounded system. The term closed is being used in a similar fashion to the way the term is used in mathematics. It is meant to imply that the interactions that occur within the system work in such a way that the only thing that they produce that might be of interest to entities outside the system is a constant value. Bounded implies that things that impact the system on a continuous non-cyclical basis are to be considered within the entity we are calling a system.

The usefulness is derived from the fact that so many systems can be shown to be closed and bounded and possess a cycle time if they are properly defined. For example, an organization with its customers and suppliers is a closed bounded system and can be evaluated as such. Many things that we want to model are closed bounded systems if the model includes all the subsystems that interact within a system.

An example of this might be the air transportation system. If you address its suppliers, its customers, weather, its ground and air access as well as individual air terminals, maintenance functions and its employees within your system, it may be shown to be closed and bounded. If you leave any of these out when you try to evaluate the system, you often come up with an unresponsive environment prone to near misses; long delays for customers; impossible ground access; wasted fuel because of stacked aircraft. Any air traveler will tell you the above description can easily be applied to occurrences at many of the major airports.

However, it is obvious that a system such as the air transportation system is so complex that it would be impossible to handle as an entity. The solution is to decompose it into subsystems and modules that can be tested independently and then put back together to reflect the entire system.

SYSTEMS, SUBSYSTEMS, MODULES

It is easy to see that although what happens at Kennedy Airport within a fifteen minute period may well affect O'Hare Airport one hour later, each can be dealt with independently, provided, of course, that the information on each is available for the other. The question is: should we consider O'Hare a subsystem of the air transportation system, or is it a module of the system? Or even more important, how do you decompose this type of complex system so that a model would be testable, shown to reflect real life and yet be practical to construct?

Returning to the concept of an organization as a system: organizations often have difficulty meeting their goals and objectives when new ones are added. Certain types of objectives can make the existing organizational structure unwieldy. The stucture of the organization is not often altered before the attempt to change is implemented. This leads to frustrations for the employees and a failure to achieve the objective without even a knowledge of what is causing the failure. Employees are often thought of as obstructionists, and the failure is blamed on their resistance to change. No one can deny that there is some fear of the unknown and people do hesitate to make changes, but often this is not the major factor causing the failure. Often the organizational structure is such that the coordination needed to meet the new objective does not exist.

One of the aims of the analysis technique developed in this book is to be able to decompose a system so that the proper coordination relationships can be identified. If this analysis is done prior to trying to implement a new objective, then the organization management can decide to either reorganize or to modify their objective, or to take some intermediate step that will raise the probability of reaching their objective.

II
WHAT IS A SYSTEM? A SUBSYSTEM?

Before one can establish criteria and guidelines for system design, it is essential that when a term is used, the term has the same meaning for everyone. Thus, throughout the book, definitions will be given for terms that are in common use. As was stated in the introduction, this book deals only with closed bounded systems. Although it has been proposed that there are several types of systems similar in construct to sets, it simplifies any attempt to develop uniform guidelines and theorems on systems if you limit yourself to a specific type of system. Since many systems can be shown to be entities that possess a cycle time, and these are usually most suitable for analytic evaluation, the term system throughout the book will mean this type of system where we will define *entity* as: anything that can be evaluated as a whole; and a *system* as: the total entity to be evaluated. This must include all the people and things that interact within the system. All forms, reports and filing associated with the system (whether automated or not); and, in the case of automated systems, the programs. It must contain a structure that allows for one controlled entry to the system, and one exit from the system and for instant interactions. The system must be capable of operating through one complete cycle on the entry information.

Looking a little more closely at this definition, one can give examples of what can be considered a system and what cannot. For example, a human exclusive of his or her environment by this definition cannot be considered a system. If his or her environment is included as part of the bounded interacting system you are trying to evaluate, then humans can be a part of the system. If you wanted to evaluate a human as a system with this definition, then you would have to place him or her in a controlled environment and limit your cycle time to the smallest increment of time that the person receives information, or needed nurture from his or her environment. However, one could look at a human within his physical, economic, social and political environment and study humans as a part of a system. This approach might be helpful to gain insights about humans.

WHAT IS A SYSTEM? A SUBSYSTEM?

To simulate a forest, you must include not only the plants but the animals, the weather, and the geology in your system. To establish an accounts receivable system, you must include the customers, the operators, the reports and records, as well as the needs of secondary users of the information into your analysis of the system.

If one can accept this definition, then criteria to evaluate whether or not you are truly looking at a system can be established. For example, the following criteria can be used to test whether or not a system is: cyclic, single entry, single call, and interacting.

Single Entry

What information is needed by this system to operate?

1. Can it always be input to the system once at the beginning of a complete cycle of the system?

2. Can the system test for errors or exceptions in the input information?

3. Can these errors or exceptions be handled at the end of a complete cycle of the system without negatively impacting this system or any other system that must interact with this system?

4. If item 3 is answered negatively, then can the errors and exceptions be isolated by the system, the system terminated, and the system restarted with correct information without destroying the objectives and goals of processing the system?

Single Exit:

What information will be output by this system?

1. Can all the information output from this system wait until one complete cycle of the entire system is completed?

2. If the system cannot wait for output at the end of the cycle, can the system be restarted from the entry point without destroying the goals and objectives of the system?

3. Is all the information output from this system in constant form, agreed upon by all other systems that must use the information, and timely and accurate for this system, as well as all other systems that need the information?

Interacting:

Are the needs and limitations, capabilities and interactions of all the people, equipment and things in the systems being considered within the system?

If we assume equipment handles redundant, precise tasks better than people can, and people can handle exceptions, errors and situations that need value judgments better than equipment, then some basic criteria for testing interaction within the system can be established. If, in addition, it is agreed that all people in the system will occasionally make errors, all equipment will at some point fail, and abnormal events will occur in any system, then additional guidelines for design can be established.

1. For systems that contain people, are value judgments being left to humans? Has the system been designed so that errors and exceptions can be handled without destroying the responsiveness of the system?

2. For systems that contain equipment, are redundant, precisely calculable tasks or boring tasks being assigned to equipment?

3. Are proper backups being included in the system that take into account: record backup, equipment failures, documentation needs caused by people leaving the system, media losses, a probabilistic occurrence of errors and abnormal events.

4. Do all people who interact within the system have proper access as well as redress?

SUMMARY

A system is defined by interrelated key words.

1. *Cyclic Single Entry*—All interactions that affect the system occur within the system. All information and occurrences that affect the system from outside the system can be input to the system once at the beginning of a complete cycle.

2. *Cyclic Single Exit*—A complete cycle of the system produces internal interactions and any output from the system can be expressed as constants at the end of the cycle of the system.

3. *Interacting*—All occurrences within the boundaries of a system are interacting occurrences and no occurrence from without the system interacts with the system during one complete cycle of the system.

Once this definition is accepted, general guidelines for evaluating whether or not you are looking at a system can be established.

Decomposition of a System:

Once it is established that the entity you are looking at is truly a system, then it often becomes apparent that the entity is too large to evaluate as a whole. The systems that tend to be both the most useful and the most interesting also tend to be complex interacting entities which cannot be easily conceptualized as a whole. The obvious answer is to analyze the system so that if it is possible you can work with smaller entities that may be contained within the system.

It has been recognized that, if the analysis can decompose a system in such a way that the subsystems are independently testable, then the system itself attains several desirable attributes. It is easier to test, it is easier to modify and, finally, its relationship to reality is easier to validate. The easiest way to assure a design which guarantees this type of decomposition is to define a subsystem in the following way.

Definition

A *proper subsystem* of a system must itself be a system.

Then one can prove some theorems about systems which can be used as guidelines to evaluate if you are decomposing a system into proper subsystems.

Theorem 1

A subsystem of a system must have a cycle time that is less than or equal to the cycle time of the system.

Proof: The way we are going to prove this is to demonstrate:

1. That there exists a cycle time that produces all the occurrences and interactions in the system. (This statement must be true if you have accepted the definition of system.)

2. All occurrences and interactions of a subsystem occur within the subsystem. (This statement is true because of the definition of a subsystem and system.)

Then we will make the following assumption and prove it must be false.

Assume the cycle time of a subsystem is greater than the cycle time of the system it belongs to. Then there must be occurrences and interactions that occur within the subsystem that do not occur within the system that the subsystem belongs to because of the two statements demonstrated above.

Thus, this assumption leads to a contradiction of the definition of a system.

What we want to know now is that a system and its subsystems are of necessity hierarchies. Here we will define *hierarchy* as: A group of systems or modules organized into orders or ranks, each order or rank subordinate to the one above it; and *level* as: the hierarchical placement of a module or subsystem.

At this point, it would be desirable to establish a schematic representation so that we can picture systems. Since we still have not attempted to define or prove what goes on inside a system that is not decomposable into subsystems, we will reserve two-dimensional hierarchical representation to this aspect of system evaluation. Thus, the usual branched hierarchical chart is not recommended. Instead, let's represent a system with a circle meant to represent a sphere. The reasoning behind this choice is that it is desirable to be able to indicate that a system has at least one more dimension than any other entity, and show the subsystems within a system as non-overlapping but connected circles. The added dimension is time, and this line of reasoning will become clearer in the next chapter.

Theorem 2

All systems that can be decomposed into subsystems can be hierarchically ordered.

Proof: To demonstrate this we must show that each subsystem must be totally contained within the next higher subsystem and the system itself. Keep in mind the key words from the definition of system: one entry, one exit, interacting, and possessing a cycle time. Theorem 1 establishes the rule that each lower level subsystem must have a cycle time that is less than or equal to its parent subsystem.

Now we must establish both that each subsystem is wholly contained within its parent system and no two or more subsystems that are wholly contained within a system can have any portion of the subsystem in common with any other subsystem.

Schematically shown, we must first prove that the following cannot occur:

WHAT IS A SYSTEM? A SUBSYSTEM?

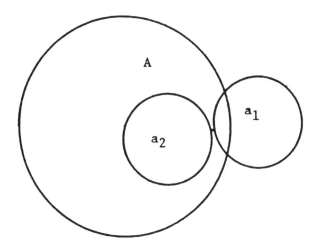

Where a_1 and a_2 are subsystems of system A, and either a_1 and/or a_2 are not wholly contained in A.

Then we must show that the following cannot occur:

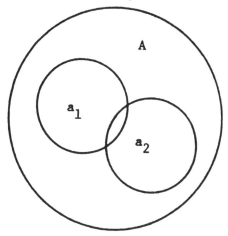

Where a_1 and a_2 are subsystems of system A and a_1 and a_2 have a portion of their subsystems in common.

First, let us assume that a_1 and a_2 are subsystems of system A, and either a_1 and/or a_2 are not wholly contained in A. Then there must be occurrences and interactions which occur in either a_1 and/or a_2 that do not occur in A. A is defined as a system and hence must be able to operate through one complete cycle on entry information alone. By theorem 1, all subsystems of A must have cycle times that are less than or equal to A. Thus, all interactions and occurrences in a_1 and a_2 must occur during the cycle time of A. Thus, either A is not a system (a contradiction) or all interactions and occurrences of a_1 and a_2 must occur within the bounds of A.

Now if we assume that the subsystem of a subsystem is also a subsystem of another subsystem on the same level, then there must be occurrences and interactions in the two subsystems on the same level which they share in common, and other interactions and occurrences which do not occur in one or the other. Since the only interaction between systems can occur through exit and entry information, then either the lower level subsystem can be broken into two subsystems, each one of which is wholly contained in the subsystem on the next level, or the lower level subsystem is not a system.

In either case the hierarchical placement is demonstrated. Keep in mind we are talking about subsystems which are defined simply as a system with all the attributes of a system. We are not talking about functions or processes. These will be discussed in the next chapter. For example, we are not trying to state that if two subsystems of a system use a sine function or calculate distance walked by the same algorithm that they are not subsystems or systems.

Thus, a schematic of a system and its subsystems might look like this:

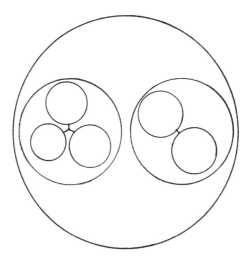

The next chapter will discuss the coordination of the internal structure of a system, but first let's illustrate where it would be applicable to

WHAT IS A SYSTEM? A SUBSYSTEM?

use what has been demonstrated about systems earlier in this chapter. When upper management wants to change management styles, and sets new goals, then since this change to the organization occurs at the top of the hierarchy, it will normally affect interactions throughout the entire organization. If this kind of change is viewed in this way, and the organizational structure is analyzed prior to implementation of the change, then there is more chance of smoothly effecting the change.

Currently, this is rarely done, and new goals are superimposed on old organizational structures. Everyone seems surprised that often a goal is not achieved and, even if it is, the process of attaining achievement is often accusatory and painful. Just this preliminary analysis of the organization as a system would help alleviate this problem. Usually the change is more permeating, so a more complete discussion of this problem will be given in the next chapter. However, a simple, commonplace example can be given here.

Assume that an organization decides it needs an integrated personnel and payroll information system. The current structure of the organization is illustrated in Figure 1.

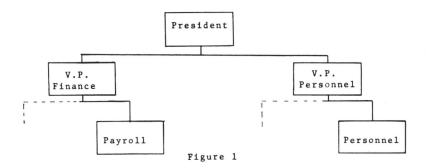

Figure 1

Assume also that there is a delegation of authority and each of the Vice Presidents coordinates a subsystem of the organization. The new objective, an integrated payroll/personnel information system, necessitates the interaction of functions that are contained within each of the two subsystems. The coordination needed to make sure that all information on the system is timely does not exist. The consequence is that under these conditions one rarely gets a system that works.

Similar problems develop in trying to model a system. Simulations and models are not the panaceas that they were once touted to be, but they are useful in gaining insights into systems. Their usefulness does depend on whether or not you are actually looking at the entire system. Frequently, the ability to decompose complex systems into proper subsystems becomes a crucial factor in whether or not you can construct the model.

A model that is useful but often turns into a bad predictor is the weather system model. This model has steadily improved because of the availability of more complete data from satellites and the increasing ability to incorporate more of the variables that affect the outcome. Satellite data has to a large extent eliminated the need to coordinate data gathering and data transfer from a good many places on the globe. Since countries are at different levels of technological development, and political considerations can interfere with data gathering, satellite capability has eliminated the need for the enforcement of universal data gathering standards. The significance of the lack of standards for information transfer will be illustrated more fully in Chapter IV.

It would be more helpful in both cases discussed above to have guidelines for even a finer breakdown of the internal coordination of the system and for information transfer. The next two chapters discuss these subjects.

SUMMARY

If we look at only closed bounded systems, and state that all subsystems must themselves be systems, then we can develop guidelines for establishing the boundaries and interactions of systems. In addition, we can demonstrate that subsystems can be hierarchically ordered and have cycle times that are less than, or equal to, the cycle times of the next level system.

The value of knowing this is that it allows you to look at the structure of an organization or a model and ascertain certain facts about it. For example, if an organization is having problems meeting a new goal, then an analysis of the organization as a system allows you to first ascertain if you are looking at the total system. In addition, you can then look at the internal structure of the organization and see whether or not the new goal has changed the interactions within the organization so what once were viable subsystems of the organization are no longer true subsystems.

Once this is done, you can either modify the goal or the organizational stucture. The judgement for what course of action should be taken would, of course, depend on other considerations also, such as a cost/benefit analysis, skills of employees, training needs, and the overall insights of the management of the organization.

III
INTERNAL STRUCTURE OF SYSTEMS

Not all systems can be decomposed into subsystems. This does not mean that the structure of the interacting processes within the system are not complex. It simply means that the nature of the system is such that within the system there is always continuous interaction. Stated another way, all entities within the system receive continuous information from at least one other entity.

It is desirable to be able to structure these systems and to produce guidelines for them also. As a matter of fact, even systems decomposable into proper subsystems need some internal structure even if it is simply to structure the orchestration of the interaction of the subsystems and the coordination of entry and exit information.

In order to achieve these guidelines and rules for internal structure, we will need some more definitions. The first thing we will define is a module of a system in which a module, unlike a subsystem, can have continuous interaction with other units of the system.

Module: Is an identifiable entity that occurs within a system. It can be an entity that coordinated the actions of other modules, or subsystems within the system, or it can accomplish a process. It may or may not possess a cycle time. Hence, there can be continuous input and/or output with other units of the system. Thus, there are two kinds of modules; each is defined below:

1. *Coordinator Module:* Is a module within a system which has the following attributes:

    1. a. Can evoke *subsystems or modules* within the system.
        b. Can pass information directly from one subsystem or module to another within the system, or place INFO values (see definition below) in a permanent record of the system.
        c. Make decisions based on flags or switches (see definition below) set by other modules or subsystems within the system or by entry information.

INTERNAL STRUCTURE OF SYSTEMS

2. *Process Module:* Is the module within a system which has the following attributes:
   a. Calculates an INFO (see definition below) value.
   b. Performs an iterative process.
   c. Places a value in an INFO that will be used as a flag or a switch.

Information (INFO): An INFO is a unique member of a system that has specific attributes for this system. The INFO's attributes can be defined by:

1. A constant
2. A decision table
3. An attribute table
4. A logical relationship
5. A relational relationship
6. A probabilistic relationship

INFO Attribute: Is the characteristic that (uniquely) defines the items of a system.

Switch or a Flag: An INFO that will be used by the coordinator to determine the flow of a system.

Based on these definitions, we can now begin to develop more detailed theorems that relate to modules, subsystems, and systems. At the end of this and the next chapter are illustrative examples of the application of each of these theorems. The reader who wants to gain insights into how each of the theorems can be applied before reading the proofs can skip ahead to page 30.

Theorem 3

If the system contains only one module, it must be a coordinator module.

Proof: Since by definition a system is the cyclic interaction of functions, then if one assumes that the only module it contains is a process module, one can show it is not a system. The definition of *process module* restricts this module to simply establishing the state of information. It cannot coordinate any interaction within the system, or pass information within the system or outside the system. Thus, if the only module of a system was a process module, you would have a process or a function and not a system. This theorem allows us to begin to show what the relationship between subsystems and modules of the system are.

Theorem 4

A proper subsystem of a system can be a coordinator module of the system but it cannot be a process module of the system.
Proof: By definition, a subsystem is itself a system. By theorem 3, if it only contains one module, it must be a coordinator module. Hence, a subsystem cannot be a process module.

We can now attempt to answer a question that was posed in Chapter I. Is Kennedy Airport a subsystem of an air transportation system? If it is a subsystem, then by definition it is also a system.

If one means by Kennedy Airport the portion of the airport that is run by the Port Authority of New York and New Jersey, and does not include the air controllers or weather station, then one can see it is a process module, and by the theorems demonstrated, it is not a system and hence not a subsystem of the air transportation system.

This is a bit difficult to accept without proof, so let's inspect this statement more carefully. The Port Authority of New York and New Jersey certainly do make decisions about Kennedy Airport. They decide on maintenance, on fees, on expansion, etc.; so, Kennedy Airport, as part of the Port Authority of New York and New Jersey system, is certainly not a process module because the chief operating officer of Kennedy can coordinate these items, schedule maintenance of the airport, and levy the fees that the Port Authority determines. However, we are not discussing the airport as part of an airport maintenance system, but as part of the air transportation system. Assume for a moment that one day the Port Authority disappeared, and all that remained in place were some rules for maintenance of the facility and the scheduling of gates and baggage handling when planes landed. Since the airlines and the FAA determine flights in and out, the surrounding states and local communities determine the access roads, the air controllers determine landing and takeoff patterns, and the airline companies handle their cargo and their maintenance, then the airport itself in this system simply processes this information and follows some well-defined rules that produce continuous output for the system.

This is not meant to imply that a change at Kennedy Airport would not affect the air transportation system. Any change in an important module of a system would change the system. It just begins to point out that the agencies who control our airports are separate, unrelated, and noncoordinated systems from the air transportation system itself. Their

INTERNAL STRUCTURE OF SYSTEMS 19

facilities are used to process a phase of our air transportation system and unfortunately are often a bottleneck in the system.

The same thing happens in organizations. The best example of a critical process module that can affect the system, but is not a proper subsystem, is the add-on computer center. This will be explored more fully later in this chapter.

At this point, let's pause again so that we can set some way of depicting these two types of modules schematically.

A *process module* will be depicted by a rectangle.

```
┌─────────┐
│ Process │
│ Module  │
└─────────┘
```

A *coordinator module* will be a rectangle with two curved sides.

```
⎛ Coordinator ⎞
⎝   Module    ⎠
```

Next to be investigated are the connection between these modules and subsystems, and the internal structure of a system.

Theorem 5

An INFO value can only be calculated in one, and only one, module of the system. For all other modules of the system, an INFO must be a constant.

Proof: From the definition of INFO, it is a unique member of the system. Assume that an INFO's constant value can be calculated in more than one module. Then, at any time during the cycle of the system, the INFO could have more than one value for the system. This is a contradiction of the definition of INFO.

A *corollary* of this theorem is that an input INFO can only be recalculated by one module of the system. The proof of this comes directly from the definition of an INFO and theorem 5.

Think back to the integrated personnel/payroll system we began to discuss in Chapter II. This theorem begins to point out why there is often trouble in establishing such a system without first instituting organizational changes. Payroll and personnel files have common data elements. Some INFOs on both personnel and payroll records must agree and are used by both offices. If the cycle of updating is different for the two subsystems responsible for the integrated information system, trouble and finger pointing result. After the proof of the next theorem, I will return to this discussion.

Before we get to structure, we need to define a new term.

Branch: A unique path from the top of a system to the end of a two-way path which ends in a module of the system.

Now let's begin to examine internal structure of a system or a subsystem.

Theorem 6

There must be *one* first level coordinator module for a system that can be decomposed almost entirely into subsystems.

Proof: From the definition of a system, there can only be one entry to the system at the beginning of the cycle of the system. If the module that controls that entry is a process module, it cannot transfer the information to any other module in the system. Hence entry information must be controlled by a coordinator module, and it must be controlled at the first level before any processing of information begins.

Assume there can be multiple first level coordinator modules of a system that can be entirely decomposed into subsystems. Since the definition of a system excludes a process where there is no interaction and a coordinator module is the only evoker of a subsystem, then from the definition of a system and of an INFO, and from theorem 5, there will be periods of time where there is no interaction between the multiple subsystems on the first level unless there is one coordinator module to evoke all the subsystems. Thus, there must be *one* first level coordinator of a system that can be decomposed almost entirely into subsystems.

Again if you look at the integrated personnel/payroll system we have been discussing, you can see that retaining the old organization structures in effect results in having two first level coordinators for one system. What, of course, results is no coordination and hence the finger pointing mentioned before.

The example in Chapter II for weather systems illustrates a similar type of problem. The information to be gathered was to be used in a system, yet there were multiple first level coordinators of the information prior to the availability of weather satellites.

INTERNAL STRUCTURE OF SYSTEMS

The next few theorems address the opposite situation in which there is only one coordinator in the system. This situation exists in any organization or system run along dictatorial lines. If the manager of a local supermarket cannot decide what items are to be in the store's inventory, then he or she is not coordinating a subsystem of a system. In spite of this fact, the solution to falling profits is often to fire the manager instead of trying to find out why things are not working. Organizations have gone bankrupt using this approach.

Theorem 7

If there is only one coordinator module in the system, the system has *no* subsystem and, conversely, if there are subsystems, then the system must have more than one coordinator module.

Proof: Let's look at two cases. First case, the system with only one module, by theorem 3, must be a coordinator module. Since a subsystem must itself be a system, then the only subsystem a system with one module can contain is the coordinator itself. Thus, the system and the subsystem must be identical.

Case two—the system with several modules. Since there is only one coordinator module, then all other modules are process modules. None of the process modules can be a subsystem by theorem 4. Assume there exists a subsystem, then by theorem 5, if there existed a process module outside the subsystem formed by the one coordinator module, and the other process modules, then there would be no way for the information to be transferred within the system, as only a coordinator module can transfer information by definition. Therefore, all process modules would have to be within the assumed subsystem, and the subsystem and the system would be identical.

Schematically, we have shown that you simply *cannot* have a system that looks like this:

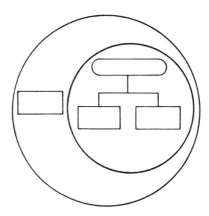

Now, conversely, we can show that any system that can be subdivided into subsystems must have more than one coordinator module in the system. The proof of this depends on the way that we have defined a subsystem as a system and theorem 6.

Along this same line, we can prove that the system itself must be hierarchically ordered and if it contains more than one coordinator module, it may be decomposable into subsystems.

First, let's look at the decomposition into subsystems.

Theorem 8

If a system contains more than one coordinator module, it may be decomposable into subsystems.

Proof: By theorem 6, there can only be one first level coordinator module for a system that can be decomposed into subsystems. By definition, a subsystem is a system. By theorem 7, if there are subsystems, then there is more than one coordinator module in the system. If follows then that systems with multiple coordinator modules may be decomposable into a number of subsystems.

We can now demonstrate that a system with n coordinator modules will have at most *n-1* subsystems.

Theorem 9

A system with n coordinator modules will have no more than $n-1$ subsystems.

Proof: First let's look at the case where there is only one (1) coordinator module. Then by theorem 7, there are zero (0) subsystems.

Now let's look at the more general case where there are n coordinator modules. Theorem 7 tells us that the first level coordinator module is the system coordinator. Thus, there can be no more than $n-1$ subsystems.

Assume there are less than $n-1$ subsystems. Then there exists at least one coordinator module that is not a system. Since we have shown in theorem 3 and 4 that if a system contains only one module, it is a coordinator module and a coordinator module can be a system.

If these coordinator modules coordinate the interaction of process modules, then by comparing the definition of process modules, coordinator modules, and systems, it is clear that all the attributes of a system are contained in the combination of coordinator modules and the process and coordinator modules that they coordinate with one exception. It is unnecessary for a coordinator module and/or process modules to have a cycle time—they can receive continuous information from the system and each other. Thus, one can state that a coordinator module and/or a coordinator module and its associated process modules are a system if the entity has a cycle time (this will be our next theorem). Thus, we have shown that there are at most $n-1$ subsystems, and from the argument above, can state the following.

Theorem 10

A coordinator module and/or a coordinator module and its associated process modules are a system if and only if the entity has a cycle time.
Proof: See theorem 9.

Therefore, it is possible that one or more of the subsystems of a system contains coordinator modules and their associated process modules. Before we attempt to draw this schematically, let's show that the system itself must be hierarchically ordered.
Schematics will fit into this proof easily as examples.

Theorem 11

All systems that contain only one coordinator module are hierarchically ordered.
Proof: Let's examine the system with only one coordinator module. Then, by the definition of process module and coordinator module, and by theorem 3, the other modules must be process modules or there are no other modules in the system. If they contain process modules then to have access to input information the modules must be coordinated by the coordinator module, cannot access each other or trans-

mit information and, hence, must be a level below the coordinator module in the system or they cannot be part of the system.

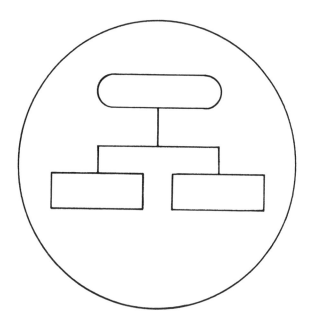

Theorem 12

All systems that contain only one first level coordinator module with its associated process modules and subsystems are hierarchically ordered.

Proof: Let's look at the more complex system where there are subsystems. By theorem 2, the subsystems themselves must be hierarchically ordered. Theorem 11 tells us the process modules must be hierarchically ordered. Thus the system must be hierarchically ordered.

Schematically, the system then might look like this:

INTERNAL STRUCTURE OF SYSTEMS 25

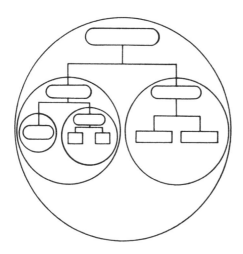

Let's pause now and review what has been demonstrated and relate this to systems that we know.

We have shown from the few basic definitions we can prove that some systems must be hierarchical. That within these systems, subsystems can exist and be well defined. That subsystems either are a coordinator and some processes, or they must by the nature of systems themselves be subdividable into other coordinators and their associated processes. This, in itself, is very significant. Let's look at an organization and its customers and suppliers for a moment. This, by the definitions we have been using, is a system. The organization will have to coordinate many processes and depending on size, objective, and how authority is delegated, will determine its structure. Classical structures of organization have come into question when organizations add new goals and objectives, or add new process modules to the organization.

It would be helpful here to further illustrate what these theorems imply for an organization. Let's return to the statement at the beginning of this chapter about add-on computer centers. When organizations first began to use computers, where to place them in the organizational structure was a problem. The solution was often under the vice president for financial affairs, or the chief fiscal officer of the organization, probably because computers were expensive, and the first applications were often accounting procedures. The equipment and the fast building staffs to man the equipment were too expensive a process module of the organization for the computer to remain a glorified accounting machine. To justify its existence, the applications began to diversify, and larger software design and development staffs were added to the centers.

The processes that were automated began to cross organizational lines. Customer complaints started, and finger pointing among employees, if not already established, blossomed. The solutions varied. Often it was: fire the computer center director. Another popular solution was to alter the organizational structure and put the computer center directly under the president or the chief officer of the organization. But unless the organization's structure was carefully analyzed, the frustrations remained.

The problem with simply leaving the computer as process module for the organization, and placing this under the chief executive office, is that you often get a system that looks like this:

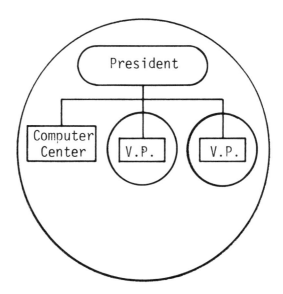

Where there are proper subsystems, the chief executive officer of the organization must be the coordinating module for these subsystems and the computer center processing. If the organization is very small this might work because all the functions might be simply process modules coordinated by the chief executive officer. Usually there are other subsystems within the organization, and a chief executive officer does not see this coordination as his or her role in an organization no

INTERNAL STRUCTURE OF SYSTEMS

more than he or she wants to be directly concerned with the day-to-day problems of the heating processes of the organization.

Again, as we explained in the air transportation system, the computer center itself is making internal decisions on how to operate and maintain both hardware and software, and on how to design systems. Changes will affect the organization in the sense that a poorly managed center can destroy the organization, just as a poorly run boiler can blow up the building; but essentially, if in the organization system, if the center director has no policy level power and simply functions as an add-on service for the organization, the computer center is a process module. Once its rules are set, it simply processes information coming from the organization on a continuous basis.

If one complicates the matter further, by adding a new goal such as planning information, necessitating the crossing of organizational lines to coordinate the information, the system might get even more unwieldy. For example, let's look at a classic organization of a college with a computer center as a process module.

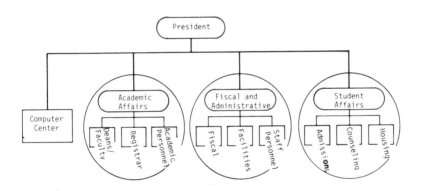

Planning information for a college comes from admissions, course, student, alumni, fiscal, personnel and facility records.

In the chart above, the only coordinator of all these record-keeping functions is the chief executive officer. He or she is the person who wants to be supplied with this information for planning purposes—they usually do not see as their role the coordination of maintaining the information. So again problems and frustrations develop, but the objective is not achieved.

Let's look at a totally different type of endeavor with systems. If you are going to simulate a system, the theory says that you better understand the boundaries of the system. Current economic models have been terrible predictors. This includes simple models that are not ambitiously trying to predict the state of the nation's economy, but more modestly the cost of doing something. Look at the model that tried to predict how much federal privacy legislation was going to cost to implement. The theory developed here says that unless you include all the interacting factors in your system, you do not have a system. Well, cost models such as the one cited above tend to ignore the return of benefits to the organization and as a consequence tend to overestimate the costs, as actually happened with the federal privacy implementation cost model.

Models then become a tool to get your point across whether your point of view has any basis in fact. Actually you can do the same thing with statistics since most people do not understand statistics any better than systems.

What has been demonstrated here is that once one clearly defines one's terms, a clear understanding of the stucture of systems can be developed.

In the final chapters of this book we will look at an organization and a simulation model in some detail. Although we know a good deal more about system structure than we did before, we still have not shown how information should be transferred. Without these guidelines, it is difficult to completely relate system structure to real life or to discuss what level in the system a particular subsystem or module should be implemented. So what we have presented in this chapter are the problems, not the solutions.

This chapter ends with summary statements for all theorems developed so far, so that they can be easily referenced in the next chapter and illustrative examples that show each theorem's relationship to occurences and problems within organizational structures.

Theorem Summary Statements:

Theorem	Summary Statement	Page
1.	Subsystem cycle time is less than or equal to system cycle time.	9
2.	The subsystems of systems are hierarchically ordered.	10
3.	A system with one module, the module is a coordinator module.	17

Theorem	Summary Statement (continued)	Page
4.	A coordinator module can be a subsystem but a process module cannot.	18
5.	An INFO value can only be calculated in one module of the system. For all other modules, the INFO must be constant. Corollary: an input item that is a constant can only be recalculated once by the system.	19
6.	There must be one first level coordinator module for a system made up of subsystems.	20
7.	A system with one coordinator module has no subsystems and, conversely, if there are subsystems, there is more than one coordinator module.	21
8.	If a system contains more than one coordinator module, it may be decomposable into subsystems.	22
9.	A system with n coordinator modules will have no more than $n-1$ subsystems.	22
10.	A coordinator module and/or a coordinator module and its associated process modules are a system if, and only if, they have a cycle time.	23
11.	All systems that contain one coordinator module are hierarchically ordered.	23
12.	All systems that contain one first level coordinator module with its associated process modules and subsystems are hierarchically ordered.	24

The goal of this part of the chapter summary is to show how rules (i.e., theorems) developed objectively and shown to be true for all systems can be applied to analyze man-made systems. In the summary of the next chapter I'll use a model of a system that contains a naturally occurring entity as part of the system to be analyzed.

There is nothing new or startling about any of the rules developed.

Any good analyst who has the task of advising on organizational structure, or the implementation of an automated process or management information system, has probably used similar rules as a guideline to his or her analysis. The only advantage of these rules is that they can be shown to have been developed independently of the personalities involved in an organization and are applicable to all systems of the type defined. Hopefully, this will help avoid the finger pointing that often accompanies problem solution in an organization.

Since they are developed for the general case, and individuals and skills do affect whether or not a particular solution will work for an organization, it is important that they be applied with that in mind. As you will see in Chapter V when I go through an analysis of an organization, there is never a unique solution if these guidelines are being applied.

Solutions for the problems that beset organizations are more dependent on their particular goals and objectives than on their structure. Of course there are structures that are so unwieldy and inappropriate for the situation, that they contribute heavily to the failure to produce solutions for problems.

So, although there are no unique solutions, there are situations that can be identified as unsuitable. In the review of the theorems presented below, an attempt is made to illustrate each theorem with situations that can be identified as problem producers.

Theorem 1

Theorem 1 points out that the overall cycle time of a system with subsystems must be greater or equal to the cycle times of its subsystem. If you look at an organization where there is a central office and branch offices that are supposed to operate as separate and independent cost centers of the organizations, it is easy to see how a failure to adhere to this rule can cause problems.

The central office should be setting objectives and goals for the entire system. There should be enough lead time in the planning so that the branch offices can rethink their goals and objectives if necessary. The central office also needs enough time to evaluate the plans of the branch office and offer any coordination advice that is necessary.

When this timing rule is violated, the branch office often fails to meet the goals of the central office. If too many branch offices are having the same problem, the organization can start to disintegrate.

Theorem 2

This type of organization described for theorem 1 is the type that often contains true subsystems. The branch offices (or subsidiaries) are expected to meet goals and objectives without continuous input from the central office which serves as the coordinator module for the system. The organization is hierarchically ordered and the responsibilities and accountability of the participants in the system demonstrate this hierarchical placement.

If one of the subsystems wants to operate in a different way or establishes goals and objectives that cannot be coordinated within the plans of the system coordinator, then the hierarchical structure is disturbed. When that happens, the organization leaves its "steady state."

Often the problem is resolved by the original system returning to a structure closely resembling its original state and the dissidents form a competing organization. Several of the computer manufacturing firms were formed by this type of challenge to organizational hierarchical structure.

Theorem 3

There are also organizations that are very simple in structure that when examined can be evaluated to be the coordinator module that comprises the entire system. This structure is often successful when the goals and objectives of the organization are few and easily agreed upon, and the success or failure of the organization depends entirely on the skills of the participants.

A small group of lawyers who have formed an equal partnership to practice general law would be one example of this type of organization. If the organization grows and diversifies its objectives, then the one coordinator module for the system can become an inappropriate model.

An organization that recognizes that its structure is becoming unwieldy can often avoid the problems inherent in trying to coordinate a system in an inappropriate fashion. Often the reason for problems beginning to occur in what was once a stable environment is that the impact of change is not recognized. This can work to dissolve a once successful and supportive partnership.

Theorem 4

Theorem 4 tells us that the partnership discussed as an example of theorem 3 should be careful not to set up a process module that they expect to operate as a subsystem. There are many organizations that have taken this route without realizing that the system structure is now unwieldy.

The worst cases occur when all planning and all resources are actually controlled by a higher level coordinator module and a subsidiary of the organization is expected to establish its own customer base and deal with marketing and suppliers. An example of this would be a college computer center that is expected to sell services to outside customers to offset its costs. The college administration continues to control the center's budget, it does not return a percentage of earnings to the center and it offers neither the budgetary support for the center to hire someone to market or deal with suppliers and these functions are not supported centrally.

The center can set up its own rules for operations, programming standards and security and privacy regulations. If there was enough equipment and personnel to make this project feasible to start with, then, assuming the center is managed properly, it will succeed for a short while. However, it will not take very long for a limit to be reached in furthering this objective.

Theorem 5

This rule, which is the first to deal with information (INFO), alerts us to a common problem in organizations that want information for planning purposes. Recognizing the problems associated with the timeliness and accuracy of information is often the first step to successfully implementing a procedure for obtaining the needed planning data.

This is a simple and obvious rule expressed by this theorem. Surprisingly, a great many organizations have spent endless hours trying to resolve problems that come directly from a violation of the principle of coordination expressed in this rule without realizing that this is the root of their information problems. It usually isn't Jones or Smith at fault, the problem is that Jones and Smith have been asked to complete a process using a critical item of information whose value has never been coordinated for the system.

This type of problem usually gets resolved if the organization hires an independent consultant or establishes a task force of employees who are not involved directly in producing the wanted results. I have seen the situation drag on endlessly, each party becoming more defensive and sure that the other person is at fault. This not only does not get the task done, it is disruptive to the organization far beyond the importance of the task.

The example given earlier in the chapter of the integrated payroll/personnel system is a classic place for the problem to first show up. Government reports necessitating the coordination of information from many sources also highlight the problem.

Theorem 6

In theorem 6 we return to basic structural rules of systems. What it is saying is that if you want separate entities who have their own cycle time (subsystems) to operate together as one system, then you can only have one coordinator module. Again, a simple, almost obvious, principle.

Yet we are always surprised when the United States fails to operate as a system and the federal government fails to be able to address certain types of objectives and goals. Our constitution provides for three coordinator modules, and the only time we get something close to these modules acting as one is in time of war.

The checks and balances of the three coordinator modules were of course deliberately built into the system and have been carefully maintained. This is not meant to imply that this structure for a system is necessarily bad. As a matter of fact, it has worked very well for the country so far.

However, with new national objectives and goals being imposed that appear to be as critical to survival as facing an enemy was in the past, we are beginning to question the appropriateness of the structure to meet these new goals. Obviously, the establishing of a Department of Energy has done very little to provide the leadership needed for long and short term planning for what now appear to be limited resources.

Currently we are involved in the finger pointing mentioned while discussing the last theorem. We hear that the problem can't be resolved because the President or Congress or both lack leadership ability. We know that sometimes dictatorships work to resolve this type of problem, but we have demonstrated over and over again we don't consider this an appropriate solution. Actually, if one looks at the Soviet Union, one can see that although the structure is theoretically sound, from the point of view of the rule of this theorem, they often fail to solve similar types of problems. So, although this rule is one clue to how you could get started on reorganizing the government to address the new goal, it is probably not the answer we seek or want because of conflicting goals.

Theorems 7, 8, 9

These rules can be used to help us analyze the problems organizations have when they mistake process modules for coordination modules or subsystems. As noted in the discussion of theorem 4, this mistake can lead to all sorts of problems. If it goes on long enough, productivity declines and absenteeism and turnover increase. This is caused by the frustration of employees trying to operate in an impossible environment.

Theorem 10

Theorem 10 shows that there can be systems that contain no subsystems. The U. S. government, if analyzed correctly, is probably this type of system. Although we constantly hear about attempts to produce five year plans and to manage by objectives, the management seems to be crisis management.

One of the reasons is that there is no continuity to the management. Top level posts in all agencies are filled by appointment, we elect a President every four years, representatives every two, and senators every six. Since campaigns for election have become so long, a great deal of the time between elections is spent in campaigning for re-election.

The coordination needed to review and revise long term plans on a cyclic basis does not exist. The delegation of authority needed to meet specific goals and objectives only exists for short periods of time if at all.

This factor is probably more the reason why the federal government cannot solve certain types of problems than the top level coordination discussed in the theorem 6 discussion.

Theorem 11

Theorem 11 tells us that any system with one first level coordinator module can be hierarchically ordered whether or not it contains subsystems. If your reaction is 'ho-hum', you're right. It was proved so that it could be used in subsequent theorem proofs.

Theorem 12

Shows us that there is nothing structurally wrong with a system that has subroutines and process modules all coordinated by the system coordinator module. Sometimes this is done accidentally because the role of a service is not understood. Sometimes it is done deliberately to limit the role of a process. Many affirmative action offices report directly to the president of the organization. If a president has established this office to conform with affirmative action guidelines, and not to really establish an affirmative action policy in the organization, the office is deliberately assigned the tasks of a process module.

Problems occur in organizations when a service is assigned the role of process module and it is expected to interact with other modules or subsystems in the system.

IV
INFORMATION TRANSFER AND LEVELS

The interactions in systems occur because of information transfer. The word information is used in the broadest possible sense of the word. The nutrient taken from the soil and input to a growing plant is considered to be information, or more precisely in the terminology that has been used in this book—INFO. INFO's values, at any point in time, would be determined by the original INFO's attributes. In the case of nutrients to a plant, the original INFO's attributes would be the differential equation which precisely defined the flow of the nutrients from the soil to the plant. All other INFOs determined by this INFO's attribute would be time-related and constant. If one thinks of information in these terms, then one can see that information transfer becomes a crucial factor in the proper analysis of hierarchical structure of a system.

Since the determining factor that separates a system from a module is whether or not the information transfer is or is not continuous, then the analysis of whether a subdivision of a system is a subsystem or module is also very dependent on an analysis of information transfer.

If we can show this crucial importance of information transfer for system design, the reason for some of the things we have all experienced will have been demonstrated. The importance of information transfer is recognized. An organization which has tried unsuccessfully to implement a Management Information System will often find itself in a position where the only hope of finding a solution to its problem is to go back and look at what information is being collected, how it is being used, and what time schedule it must be used on and by whom. Similarly, when models are not simulating the real life situation they are meant to represent, the first step in isolating the problem is to go back and trace information transfer paths.

In both cases, to totally solve the problem, you will have to go back and look at your organizational structure, your goals and objectives, and your hypothesis—but looking at information transfer gives you the clues as to where you have to begin. Why this happens is easy to understand if one looks at some naturally occurring systems.

Physical scientists and engineers describe the state of a closed, bounded, naturally occurring system, such as an atom, that is not being affected by outside forces as a "steady state." This is a very descriptive term because, although there are interactions within the system, the system itself does not change. Apply a significant outside force that acts as input to the closed bounded system and the atom leaves its 'steady state' for a while and reacts to the information transfer from outside the system. It will eventually return to a 'steady state' cycle time when the internal interactions caused by the input INFO is completed. However, it may return to a new 'steady state' and become a new system with new INFO attributes.

I will use the term *'steady state'* and define it to mean the condition that describes the interactions within a system uninterrupted by new entry INFO during a cycle of the system.

The same type of thing happens to organizations. The organization adds a new unit or imposes a new goal on the existing structure. New information transfer paths are needed, different interactions occur, and the organization leaves its 'steady state.' The same type of thing will occur if the attributes of either the customer base, or the organization's suppliers, is dramatically changed since they are also part of the organization's system.

The first thing we will prove about information transfer is that the level of the subsystem is determined by its need for entry information. You will find illustrative examples of the theorems in this chapter on page 43.

Theorem 13

The hierarchical placement of a subsystem is determined by where its entry INFOs are obtained.

Proof: Assume that entry information can be obtained for a system from its subsystems.

Theorem 2 states that the subsystems of systems are hierarchically ordered.

The definition of system and subsystem state that there must be a single entry of information at the beginning of the system cycle time.

Theorem 1 tells us that the cycle time of a subsystem is less than or equal to the cycle time of the system.

If the subsystem is establishing the value for an entry INFO of the system, then it is doing it during the cycle time of the system by the rules above. This contradicts the definition of system.

Since all subsystems are defined as systems, the theorem is proved.

Definitions

Global INFO: Is an INFO whose constant attributes will be used as entry information in more than one module of the system.

File: Is a related collection of global INFOs and their constant attributes.

In many organizations, the failure to understand the need for only one module of the system to set a global INFO leads to problems. Before we proceed, let us look at a situation that occurs in many organizations. Let us reexamine an organization that has a payroll system coordinated by one module of the organization, and a personnel system coordinated by a different module of the system. The grade level of an employee is often a decision value for both offices. The personnel office may use this item to determine such things as number of days of annual leave, and the payroll office may use the grade level to edit their files and check on the proper ranges of a paycheck being issued. If they maintain separate records, and each can alter this INFO value on their records, the probability of introducing errors into the system are greatly increased. If someone fails to inform payroll of a change or the change is read or recorded erroneously, the editing in payroll can fail. In the case of a demotion, it is unlikely that an employee would report errors in a payroll check to the payroll office.

This example leads us to examine the question of where and on what level and branch should a decision value (a switch or a flag) be set. The proper placement is demonstrated in the next theorem.

Theorem 14

If the value of a switch or flag is determined by an entry INFO, then the value must be set by a subsystem that is hierarchically above the first level coordinator module of the subsystem making the decision or it must be determined externally to the system.

Proof: Theorem 13 establishes the rule that the hierarchical placement of a subsystem is determined by where its input information is obtained. Since a switch or flag is only an INFO that is used for decision purposes, the theorem to be proved follows directly from theorem 13.

What happens if you violate this rule? The situation of the payroll and personnel office each needing to know the grade level of the employee described previously is the result. It is a fairly common occurrence in organizations to have information that is not timely for an operational office. This usually occurs when the information is the responsibility of an office on the same level as the office which needs the information for decision purposes, and the two functions report to different coordinators.

Now if we examine global INFO, something more about the level that switches and flags are set becomes apparent. One can show:

Theorem 15

The constant value of a global INFO must be set in a system that has a longer cycle time than the systems that are using the item as entry information.

Proof: By definition, a global INFO is an INFO whose constant value is used in more than one module of the system. By theorem 5, a global-INFO can only be set in one module of the system.

Assume the global INFO is set by a module in a subsystem that has a shorter cycle time than that of a system using the INFO as an entry-INFO. From the definition of an INFO, it must have a unique value for each cycle of the system. Since the INFO can be reset every time the shorter cycle time system is evoked, the longer cycle system can have entry INFOs more than once in a cycle. This cannot happen because of how we have defined system.

Corollary: A global INFO cannot be set during the cycle time of a subsystem using the global INFO as input or the 'steady state' of the system will be impacted.

Proof: Assume this restriction does not exist. Then the value for the global INFO that is an input item for a subsystem will change during the cycle time of the subsystem. This violates the restriction on systems that they have one entry during a cycle time. Entry information to a system during the cycle will have the effect of making the system unstable.

What implications does this theorem and corollary have for an organization? It shows why a procedure must be developed and agreed upon for the timing of updating of information for all INFOs that will be used by more than one office in an organization. Frequently, the failure to establish such procedures and obtain a commitment to the proper cycle time for updating an INFO is the direct cause of an inability to obtain planning information. It can also be the major cause of a tremendous blunder by the organization.

For example, let's look at an organization that has a centralized department that issues all the checks for an organization with many geographically scattered branch offices. Assume that rules for eligibility for receiving payment change, and the branch offices must review all their clients' status in order to update their files and transmit this information to the central office. If the cycle time in which the branch offices make the corrections is greater than the cycle time for check mailing, then there is a good chance that clients who should not be paid will receive checks.

This is what has happened in many welfare centers. The worst cases occur when the actual updating of the files is done by a centralized data and control group.

The branch operational group which has a shorter cycle time because of direct client interface is setting a global INFO that will be used as a control by a group that is not being coordinated by the branch office. If the system designed for the check issuing module does not ensure that the procedures meet the timing needs of the branch office and the central operating office, chaos results. The central office data control group can try and put pressure on the branch group for timely update but, because the only coordinator is at the top management level of the organization, it often fails to exert a timely reaction.

Does this example mean that you cannot have a central computer facility that issues all the checks for an organization? Not at all. It simply means that you must have a different organizational structure if you want to take that approach. In the next chapter, when we look at an organization, we will offer several alternate solutions.

Finally, let's look at a system's files. We can show that it is desirable to organize files around the input needs of subsystems.

Theorem 16

The number of entry files needed by the subsystems of a system can be determined by relating the placement of global INFO on a file to the branch of the system where they will be used as entry INFO.

Proof: Global INFO, by definition, is constant information that is used in more than one subsystem or module of the system. The minimum number of files containing global INFOs for a system will occur when all global INFOs are on the entry files for the system.

Assuming this is rarely the case, then let us establish when the maximum number of files will occur. If each module of the system establishes the constant value for at least one global INFO, then the maximum number of files will occur. If the system has n branches, and there are I entry files to the system that contain global INFOs, then that number will be

$I + n$

It is possible that one subsystem of a system with $I + n$ files could need $I + n - I$ files for entry information.

Thus, we have shown that the number of files is related to the number of branches of the system on which global INFOs are established. In addition, one can determine the number of entry files needed by a subsystem by relating the placement of global INFOs on a file to the branch of the system where they will be used as entry information.

Thus, the theorem is proved.

The value of knowing this is that since the system coordinator is the only module that can coordinate information between the main branches of the system, the role of the branches play in the number and type of files is important in optimizing the use of files.

This chapter concludes with a summary statement of the theorems proved in this chapter and illustrated examples for each theorem based on a model of a naturally occurring system.

Theorem	Summary Statement	Page
13.	The hierarchical placement of a subsystem of a system is determined by where its entry INFO is obtained.	37
14.	If the value of a switch or a flag is determined by an entry INFO, then the value must be set by a subsystem that is hierarchically above the first level coordinator module of the subsystem making the decision, or it must be determined externally to the system.	38
15.	The constant value of a global INFO must be set in a system that has a longer cycle time than the systems that are using the INFO as entry information.	39
16.	The number of entry files needed by the subsystems of a system can be determined by relating the placement of global INFOs on a file to the branch of the system where they will be used as INFO items.	40

Before I can get to the examples for each of these theorems, a brief discussion on the modeling of systems that contain naturally occurring phenomena is needed, since I will be using these models in the examples and I don't want to assume that everyone is familiar with them or with discrete event simulation models.

Most things we call systems that occur in nature are not the types of systems we have defined. There is no real cycle time and they are continuously receiving information and producing information to entities outside their boundaries. For example, a submarine traveling through the ocean and receiving signals from a satellite is constantly in touch with its environment and reacting to it. The plants growing in a forest are continuously affected by weather, soil conditions and animal life. However, it is possible to construct a mathematical model of the entities to be studied and the model itself can have the properties of a closed system.

The type of model that is used in this case is called a discrete event simulation. It makes the assumption that by evaluating the systems only at discrete time intervals and/or when significant events are happening, we can gain insights into the interactions of the entity. These insights can be used to test hypotheses or point out new areas where it is important to gather data.

Constructing these types of models so that they are useful tools for study has its difficulties. Yet my own and many other people's experience has shown that it can be done and useful insights are produced.

The best models are constructed in a way that makes them easy to test and modify and capable of emphasizing or de-emphasizing a particular subsystem or module. To construct a model that has these attributes, the rules established by the theorems can be used as guidelines.

I will illustrate how this applies to the problems in constructing these models in a similar fashion as in the last chapter. Further illustrations will be found in Chapter VI. Before we get to the examples, a brief discussion of the approach to modeling is provided for people unfamiliar with the technique.

Computers are the tools that have made it practical to simulate a complex interacting system. The best way to design a model that is to be run on a computer is to provide for one top level coordinator module whose job it is to judge which of the subsystems or process modules is needed at any particular moment. Models of the type we are discussing provide the most information if external information is brought in once at the beginning of the simulation, and the model is run through a predetermined period of simulated time and the output information is presented at the end of this period of time. Thus, the model design fits the definition of a system.

Within the model itself there are usually shorter periods of time that are defined by the events to be simulated. Many of the entities which people are interested in exploring are not well understood. This is one of the reasons people build these models. Another reason for using a model is that some interactions people want to study are fairly well understood but too expensive to set up for the purpose of running an experiment. A good example of this latter use of models is the simulation of a military maneuver. A maneuver is, of course, a mock battle and is itself a simulation. The military have been using discrete event simulation models for many years to augment what they learn from actual military maneuvers.

Let's look at models of this type to see what information transfer problems must be addressed so that you know how best to design the model. As in the last chapter, I will relate these to the theorems of this chapter.

Theorem 13

This rule tells you that the hierarchical placement of a subsystem of a system is determined by where its entry information is obtained. One of the design problems in constructing this type of model is to make sure that all information that will be needed to start the model running is available. The rule says this type of information must be coordinated by a first level coordinator module. This is obvious, but what about INFO that will be used by several subsystems or modules during the run of the simulation?

Let's look at a submarine encounter model. (Two submarines trying to detect each other but not be detected.) Both submarines' operations will be affected by environmental forces. The first question is, can the environment be handled as a separate subsystem since its interaction with the submarines is continuous and this violates the definition of subsystem? The answer for a discrete simulation model is, yes, because only changes in the environment will produce new interactions. So there will be discrete periods of time when the input values from the environment can be considered constant. The next question to answer is on what level in the organization of the model should this subsystem be placed? The output it produces will be used as input to the two submarine systems and our rule tells us that the transfer of information must be coordinated by the coordinator model that coordinates the interaction of these subsystems. So the hierarchical placement of the subsystems is determined.

Theorem 14

This theorem will help us determine at what level the coordinator of the subsystems described in theorem 12 must be. Since what we are proposing to study is simply how two submarines playing hide and seek and the environment interact with each other, there will only be three subsystems in the model. Thus, the answer using theorem 12 is a first level coordinator module. This coordinator module would be controlling five things: the process module that brings in all information needed to start the model, the three subsystems interactions and the module that outputs the information at the end of the simulation.

Another example that might make the importance of level of input clearer is the following: assume you are simulating a forest and preliminary analysis has told you that you can handle all the plants as a subsystem. You want to know which method of dividing up the plant world into subsystems or modules will work best. Ecologists use two methods. One divides the forest by height and evaluates interactions at each height. The bottom level is the forest floor. There are two intermediate levels, and the top level is the canopy of the trees. The second groups plants into families that have similar characteristics.

If we choose the first method then any event that affects the plant subsystem, say an attack of insects in the canopy, affects all the subdivisions we have chosen because trees extend through all four divisions. If we choose the second method then some events may have no or little effect on some of the groups. The advantages of having a model where assumptions you make can be tested by exaggerating the assumed effect and noting the overall effect on the model should be obvious. This technique can be used most effectively when not every event affects every entity in the model.

Theorem 15

Let us go back to the submarine encounter model to demonstrate the types of problems you have if you violate the timing rules for global INFO established by theorem 15 and its corollary. I stated without giving the reasoning behind the statement that the two submarines playing hide and seek should each be separate subsystems.

Obviously from the point of view of efficiency of code and storage of information in a computer this is not a good approach. Since they are both submarines and both interact with their environment they will have most things in common. In fact their only differences will be in their capabilities and the rules that govern how they operate. So if one of the goals of the model is to optimize the amount of computer storage that is used this may not be a good model design.

Since this is the type of model where you want to observe what is happening not only at discrete intervals but when an event occurs that will disturb the equilibrium of the model, it is desirable to coordinate this type of event externally to the subsystem. The event is of course something that you have assumed will disturb the 'steady state' of the situation.

In organizations you want to prevent this from happening too often. That is why changes in government regulations that occur either too rapidly or in an inappropriate place in the cycle to the organization are much more difficult to implement and cause real problems within the organizational system. In a discrete event simulation you are deliberately trying to ascertain the effect of this type of event.

If you have the two submarines in separate subsystems coordinated by a coordinator module it makes it easier to make changes to the model when assumptions are proven incorrect. For example, you will want to know if the capabilities of navigational instruments affect the outcome. This can be done most easily by assuming unequal capability and then simulating events that you assume will have an impact. The events in this case will be environmental and some environmental events will only affect one of the submarines. An example would be a localized thunderstorm where only one submarine is within the boundaries of environmental disturbance.

If this event does not seem to have the effect that the assumptions have predicted, you'll want to make changes to the characteristics of each submarine's navigational equipment to find out why. You won't want to make these changes simultaneously. It is easier to keep track of the changes and experiments if each of the submarines is handled as a subsystem.

Let's continue this discussion in the illustration for theorem 16.

Theorem 16

Theorem 16 shows how the number of input files is determined by relating the placement of global INFOs on a file to the level and branch of the system where they will be used as entry INFO. One of the things it is most crucial to control in this type of a simulation is to make sure that when you make changes to the parameters or equations (the INFO's attributes) that define a capability that these changes are reflected throughout the subsystem and are isolated from other subsystems.

The type of changes discussed in theorem 15 are commonplace in any model where there are a great many assumptions. The goal of most of these simulations is to test interactions. For example, I can grow a tree in a nursery under controlled conditions and find out a great deal about that species of tree. This provides me with data upon which I can base assumptions on how that tree will grow in a forest.

The interactions of other plants, the animals and uncontrolled environmental factors in the forest may or may not dramatically change my assumptions. I simulate the environment to obtain some insights into this and into whether or not I should be gathering other types of data.

One of the most important things to control in the model is to make sure that when I want to change my assumptions I can do this efficiently without making changes to so many files that I have created an enormous bookkeeping problem. The more I can isolate and limit the branch and level that the changes will have to be made, the easier my bookkeeping becomes.

Thus if each submarine is a separate subsystem since I know in advance I will be changing the files of each, independently handling each submarine as a separate subsystem is desirable. This points out again that the objectives and the goals of operating the system are more important to how you should structure a system than any other factor.

V
SOLUTIONS: AN ORGANIZATION AS A SYSTEM

The final chapters of this book suggest some solutions to the problems outlined in previous chapters. The first thing we will look at is an organization as a system, and then in Chapter VI we will look at a simulation of a system. Since colleges are the organizations that I understand best, I will use them as an example. Contrary to popular belief, business organizations are in many ways similar in structure to colleges. Instead of students, they have customers. In place of an admissions office, they have market reps, and the development office at a college is similar in concept to the officers of a business who must obtain venture capital for the business.

Although a college only provides services, it does have a product. The product, of course, is its alumni who help or hamper the college, both in terms of direct fiscal and political support, and in enhancing or detracting from the college's reputation by their own personal and professional lives.

In many ways a college can become a more complex structure than many businesses, since the objectives and goals can be broadened to serve a diverse customer base (i.e., students, the surrounding community, research, the state) with conflicting needs. In addition, many public colleges are often forbidden by law from deficit spending. In a bad year, they cannot go to their local banks to borrow money to tide them over until times get better. This means that, in times of dropping enrollments, planning becomes very crucial to their proper management.

One further point. Although they produce a product, and by good quality control in their system can improve the product, they cannot control the behavior of their product. Unlike an automobile dealership that can be controlled by the manufacturer whose product it markets and services, an alumnus, once he has graduated, can influence the outcome at a college much more than the college can influence his life.

The simulation we will look at is a portion of the air transportation system. It would take an entire book to look at the entire system as it is very complex and, as currently structured, does not illustrate the system design standards that we want to examine. Instead of looking at today's reality, we will briefly examine a large mythical country's air transportation system and specifically within the system, the airport subsystem. This country will have the following attributes: it is large enough in land mass so that the volume of domestic flights exceeds the volume of foreign flights. It has a federal department of transportation which acts as the coordinator that determines the size and location of all public transportation terminals. The location of airport terminals is determined by the following objectives: energy conservation, accessibility to *all* citizens of the country, safety and reliability of the mode of transportation, and cost effectiveness. This country is so mythical that neither political considerations nor graft are determining factors for terminal location.

But first, the organization as a system.

AN ORGANIZATION AS A SYSTEM

A college

Using the definitions for system, a college as a system is made up of the following parts:

1. Customers
2. Employees
3. Products
4. Suppliers
5. Physical Plant

It has a chief officer usually called a President, and can be organized in many ways. Often the policy level positions directly under the President are called Vice Presidents, and historically some line management positions that are unique to colleges use the following titles:

Dean—coordinates a number of academic departments or student services

Director—coordinates an administrative function

Bursar—accounts receivable supervisor for fees

Registrar—coordinator of student records, and manages registration, course scheduling, and general academic advisement

Variations from college to college do exist, but these titles

would be understood at any college. Colleges are often subdivided into four subsystems coordinated by the President. These are:
1. External Affairs—this can include offices such as public relations, alumni, development, sponsored research, admissions.
2. Administrative Affairs—this includes: security personnel, budget and accounts, bursar, physical plant and sometimes services such as the computer center.
3. Academic Affairs—this includes academic departments and can include some services such as AV and TV, libraries, registrar.
4. Student Affairs—this includes counseling, housing, and can include food service and recreation facility management.

There are variations on this theme. The size and complexity, as in business, will determine the number of Vice Presidents, line management officers and the placement of certain offices in a college.

In a simpler era, when colleges were more monastic in concept and did not see themselves as a resource for the community, the state or the federal government structures were simpler. Today, most colleges view themselves as having a variety of missions.

The need for planning information in this type of complex business, that is often tightly fiscally controlled, is apparent. Many colleges, like many businesses, have seen the theoretic value of management information systems. The results in trying to implement these systems have been far short of the expectations. In fact, most colleges have a difficult time collecting the information that is needed simply to meet federal and state reporting requirements.

The reason, as briefly discussed in Chapter III, is that current organizational structures often make the President the coordinator of informational needs. A possible solution to this problem, and the problem of running an institution with multiple missions is presented next.

A NEW STRUCTURE FOR A COLLEGE

Let us look at a college from the point of view of record keeping. Both for federal and state reporting purposes, and for planning purposes, the following categories of information are needed:
1. Accounting (fiscal)
2. Student
3. Course and Event Offerings
4. Grants and Contracts

5. Personnel
6. Facility
7. Alumni and Contributors
8. Admissions

Information about the student, that the college must retain, relates to: housing, admissions, fees, course and course credit and grade, and financial aid information, as well as payroll and personnel records, if the student is also an employee of the college.

Colleges frequently retain all this information about students as one record, necessitating coordination among the following offices:

Office	Coordinator
Housing	V.P. Student Affairs
Admission	V.P. External Affairs
Fees	V.P. Adminstrative Affairs
Course credit, grade	V.P. Academic Affairs
Financial Aid	V.P. Student Affairs
Payroll and Personnel	V.P. Administrative Affairs

To complicate the matter, the accuracy and timeliness of payroll and personnel information is dependent on financial aid information and course credit information. The accuracy and timeliness of personnel information is also dependent on housing and admission records.

No one wants to set up a totally unwieldy situation nor, as mentioned before, do the Presidents of any colleges see themselves as coordinators of information handling. This situation has come about for the same reason that you get a similar situation in business. Demands for information have increased, organizations have become more complex, and automation makes it all seem possible.

If one begins to take a totally new look at the situation, it becomes apparent that the marketing function of the admissions office is not tied to the record maintenance function. The only tie is that the admission counselors must submit some of the input information to the record-keeping function, and they must receive some periodic reports from the record-keeping function. Similarily, the financial aid counseling function of a financial aid office is not necessarily a part of the record-keeping process. Again, there will be input and output between any of the counselling functions and the record-keeping functions. Thus, a completely new approach to organizational structure and the establishment of subsystems can be constructed.

Before proposing a new organization of the system, one has to look at what one wants to accomplish. Any organizational structure will have some drawbacks, so the goals and objectives become particularly important prior to the system evaluation.

For this hypothetical college, the objectives will be the following:

1. Ability to perform the following missions:
   a. Quality undergraduate education for resident students.
   b. Quality undergraduate education for part-time students.
   c. Quality graduate education, basically on the master's level for part-time or full-time students.
   d. Support of research, particularly those projects which explore improving educational quality, regional resources, new technology in education, and which support the college's graduate programs.
   e. Offering of educational resources (i.e., cultural events, computer services, television productions, recreation opportunities, etc.), and non-credit training to the region.
2. Adequate planning information for the multiple missions.
3. Integration and coordination of all functions on the campus (i.e., elimination of all add-on services).
4. Enough flexibility in staffing patterns so that tenured individuals can be reasonably reassigned tasks.

To meet these goals, it would be helpful to look at what offices are currently doing.

Let's look at a standard organization chart for the hypothetical college. Organization charts are, of course, designed to indicate the channels of command. Usually this is also interpreted to mean that the flow of information is coordinated hierarchically. (see page 52).

Now let's look at the same operational offices on a function chart that shows what types of services they are performing for the organization. The services are listed under the following headings: record-keeping, reporting, resource management, counselling, teaching, research, finance, marketing, public information.

The function chart on page 54 looks at the line management offices at a typical college. The column headings have the following meaning:

Reports to: Indicates the branch of a college that, under a classical organization structure, the office would belong.

Reporter: Indicates that the office would normally have primary responsibility for issuing reports based on the information collected.

Record-Keeping: Indicates the offices would have the primary respon-

SOLUTIONS: AN ORGANIZATION AS A SYSTEM

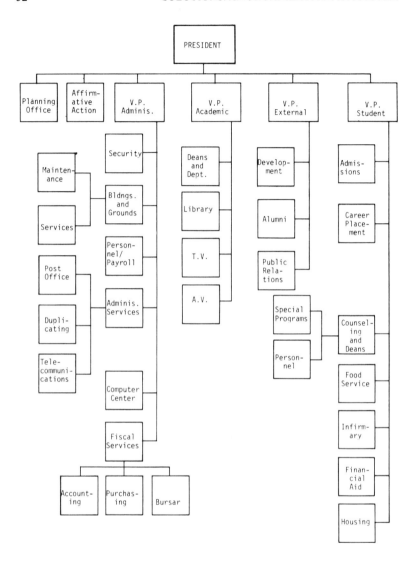

SOLUTIONS: AN ORGANIZATION AS A SYSTEM 53

sibility for maintenance of the information. Obviously, on information such as personnel leaves, all offices of the organization would be involved in data collection.

Resouce Management: The office that is responsible for maintaining equipment and/or media, and providing the service to the college.

Counselling: The aiding of people and giving advice on personal, academic or resource utilization matters.

Teaching: Instructing individuals in either the classroom and/or laboratory or resource environment.

Research: Project oriented investigations.

Finance: Collection of fees for services and the raising of funds.

Marketing: Presenting of the college's resources to individuals external to the system.

Public Relations: Presenting written and media material to individuals external to the system.

Functional analysis is one way of looking at an organization as a system, and it will allow us to apply many of the theorems of the preceding chapter to our organizational design.

Examining the function chart on page 54, it becomes apparent that many offices on the campus serve multiple functions for the college. If you look at simply those offices that are responsible for resource management, an interesting pattern emerges. Some of these offices are basically single function operations; they do some record-keeping and reporting, but are primarily responsible for the management of a resource on the campus. Some are multiple function operations and include such responsibilities as counselling, teaching, and research. The single purpose offices are usually under the administrative branch and the others are scattered among the other branches. The exception on this chart is the computer center. Placing this under the administrative branch is somewhat arbitrary. As mentioned previously, it is a fairly recent add-on and can report to almost any branch, or directly to the President, or there can be two centers, one for academic and a data processing center for administration.

Since three of the Vice Presidents are responsible for resource management functions, then coordination for campus-wide resource management can only come from the Presidential level. Checking the other functions, one sees that all cross organizational lines. We see by theorem 7 that what we have is a system with one coordinator module

SOLUTIONS: AN ORGANIZATION AS A SYSTEM

Office Name	Reports to:	Reporting	Record Keeping	Resource Management	Counseling	Teaching	Research	Finance	Marketing	Public Info.
Academic Depts. & Deans	Academic				X	X	X			
Administrative Services	Administration	X								
Post Office			X	X						
Duplicating			X	X						
Telecommunications			X	X						
Admissions	Student	X	X		X				X	
Affirmative Action	President	X	X		X					
Alumni	External	X	X						X	X
Audio Visual	Academic	X	X							
Building & Grounds	Administration	X	X	X	X	X	X			
Maintenance			X							
Service			X							
Career Placement	Student	X	X	X	X		X		X	
Computer Center	Administration	X	X							
Facility Maintenance		X		X					X	
Adm. Design & Dev.					X	X	X			
Research & Instruction		X	X		X	X	X			
Counseling Center & Deans	Student				X					
Development Office	External		X						X	X
Financial Aid	Student	X			X					
Fiscal Services	Administration		X							
Accounting		X	X					X		
Purchasing		X	X					X		
Bursar		X	X					X	X	
Food Service	Student	X	X	X						
Infirmary	Student	X	X	X	X					
Housing	Student	X	X	X	X					
Library										
Facility Maintenance	Academic	X	X	X	X	X	X			
Research Services							X			
Planning Office	President	X					X			
Public Relations	External								X	X
Personnel & Payroll	Administration	X	X		X					
Registrar	Academic	X	X		X					
Security	Administration	X	X							
Special Programs	Student	X	X	X	X	X	X			
Television	Academic	X	X	X	X					

(the President) and no subsystems. Returning to the function of resource management, and examining it more closely, we can see some problems on this approach. Since Presidents have historically been chosen from academic administration, this is not normally their area of expertise. On a campus with a single purpose or mission this might not be a problem, but on a multi-mission campus, as we are describing, conflicts arise in allocating the use of the resources among the diverse user base: students, local administration, funded researchers, off-campus service contract users, does become a problem.

Probably the greatest problem is that if the President leaves (i.e., resigns, gets ill, takes a leave), then all coordination on the campus disappears.

To resolve this type of problem and to meet the remaining objectives, a new type of organizational structure could be proposed.

Checking back on our theorems and definitions to see if a functional analysis of the organization can be used to set up proper subsystems for this organization to meet the objectives stated, we note the following:

Assuming that the proposed system design will have one first level coordinator (i.e., the President), then by theorem 11 the system can be hierarchically ordered whether or not it contains proper subsystems. If we choose a design that contains proper subsystems of the systems, then by theorems 3 and 7 each subsystem must be coordinated by at least a first level coordinator module. By theorem 1, these subsystems' cycle times must be less than or equal to the system cycle time, and by definition of system and subsystem each subsystem must be able to go through one cycle without receiving input information during the cycle.

In the September-October, 1979 issue of the *Harvard Business Review,* Elliot Jaques writes of data correlation tests he has used where he was able to show that the level of position in an organization is very highly correlated to the maximum time allowed to complete the longest task. In an article titled "Taking Time Seriously in Evaluating Jobs," he shows that both from the perspective of what the employees sense as fair pay for a job and also from the point of view of level on the hierarchical structure, time is a crucial factor.

The theorems developed point out the same thing since we have demonstrated that the upper levels of the hierarchy need a longer cycle time than any subsystem within the system to assure good planning and coordination. The role of the chief executive officer and his or her policy level subordinates for planning has been understood for a long time. The necessity for planning cycles that exceed the fiscal year is also an accepted tenet of management. The operational or line management functions of an organization must be controlled within shorter cycle times. This is also an accepted tenet of management and has resulted in things like yearly audits of financial records.

On the operational level of a college there are two basic cycle times:

1. Fiscal year—usually from July 1—June 30
2. Academic year—usually from mid-September to mid-May

Historically, faculty appointments have been made on the basis of two academic terms (30 weeks), and administrative appointments on the basis of 12 months. A multi-mission college would have to be in operation 12 months, particularly to meet the service mission, so the overall cycle time of the system would have to be 12 months. If one assumes the current academic calendar will be retained, then in a sub-system construction, teaching would have to be a subsystem of the system and coordination must be below the Presidential level.

Examining the function's chart, and checking for cycle times, one sees the following:

> Marketing has two cycle times: one is related to the academic calendar (admissions), and one related to fiscal calendar (service).
>
> Financing has two cycle times: one collection of fees is related to the academic calendar, and one to the fiscal calendar (service).
>
> Public Information has one cycle time: collecting and reporting.
>
> Information collection also has two overall cycle times for auditing purposes.
>
> Resource management has one cycle time related to fiscal year.
>
> Teaching, research and counselling has two cycle times. One that is related to the fiscal calendar and the other to the academic calendar.

Thus, if you group by functions, and develop subsystems, the subsystems coordinated by the President must have greater than a fiscal year cycle time, and subsystems coordinated by these subsystems can have a fiscal year or academic year cycle time.

Within the academic and fiscal cycle, there will have to be established shorter cycle times. Obviously for practical use, accounting information or resource utilization information cannot be transferred once a year.

SOLUTIONS: AN ORGANIZATION AS A SYSTEM

One possible system design is illustrated in the schematic.

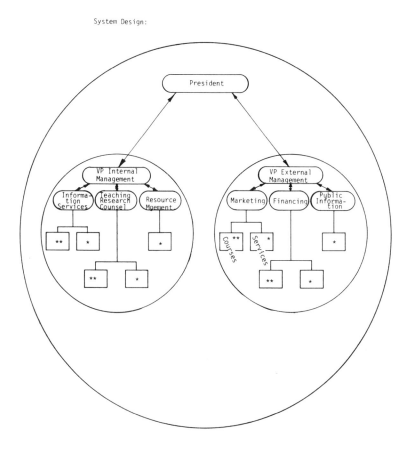

* Fiscal Year
** Academic Year

Let's examine the internal and external management structure more closely.

INTERNAL

Under resource management you would have two categories of service: learning resources (i. e., libraries, computer centers, AV, TV, communications, printing) and plant management (i. e., security, purchasing, buildings and grounds). Under academics and counselling, you would have three categories of service: counselling, research, and teaching. Here, counselling is used in the broadest sense of the word: i.e., academics, advisement, personal counseling, resource utilization counseling (libraries, television, and radio production, computer use, financial aid, etc.). Under information services you would have all record-keeping functions (i. e., accounting, student, course, etc.) and the design and development of administrative systems, as well as the Institutional Planning Office.

EXTERNAL

Would coordinate all interfaces that occur with organizations or individuals who are external to internal management, but who, because they directly interact with the organization, are part of the system.

Under marketing you would have two categories: direct marketing of teaching services (admissions) and selling of other services.

Under financing you would have two categories: development (i.e., gifts and endowments) and direct financing (fees and allocations).

Under public relations you would coordinate the official information released about the college, both to the media and by publications of the college.

Let's look at what this system design does to the current structures of most computer centers or libraries. It separates computer or library operations and maintenance of an application library (resource management) from the design and development of systems (information services) from the direct counselling of academic users that want to use the utility as an academic or counselling (academic and counselling management) tool. Most centers and libraries are presently structured somewhat in this fashion, but all three divisions are coordinated by the

SOLUTIONS: AN ORGANIZATION AS A SYSTEM

same individual. In the above structure, the three divisions would be redistributed along three separate branches of the system, leaving the computer center and the library as a utility for the rest of the system. This kind of change can only be made if the information transfers between systems are global items. In the case of a computer center, this means that all programs and automated system design must adhere to universal or organization standards.

Since computer centers are a relatively new part of the organization, this kind of change would be minor in impact compared to the change made to the classic business office. The current services in any business office (budget, purchasing, accounting, payroll, fee collection) would be redistributed radically. The following table gives you a picture of the redistribution.

Service	*Subsystem*	*Coordinator*
Budget	Internal	Information Services
Purchasing	Internal	Resource Management
Accounting	Internal	Information Services
Payroll	Internal	Information Services
Fee Collection	External	Financing

The placement of fee collection external to the internal system is important because, prior to the payment of a fee, or submission of a promissory note for a fee, last semester's student is not an official student of the college. Failure to recognize this has led to conflict situations. For planning purposes, and to establish the fee, it is desirable to have a student register for courses prior to paying the fee, but until financial obligations are met, the individual is not a student of the organization.

Tracing this design back through the theorems shows several things have been resolved. No process module is isolated and external to a system. External management and internal management can be treated as proper subsystems as each has a cycle time, and the timing for the input needed from marketing and financing to internal management and vice versa can be established by college-wide procedures. Presidents do see themselves as the coordinator of internal and external management. The interactions between information services, academic and counselling, and resource management can be coordinated at the level indicated. Each could be set up as separate subsystems with coordinator modules handling the subdivisions—or can remain in one large subsystem.

The level of the subsystems are properly placed for coordination of input to the finer division of subsystems if this is considered to be desirable.

Thus, by the theory, this would work. Planning information should be available and the college should be able to serve multiple missions of education, service to the community, the state and federal government. Would it work in practice? This is where human value judgement comes into the picture. It is a radical change from current structures that exist either in colleges or in businesses. Record-keeping has always been considered the responsibility of the department that coordinates the service, and it has been shown that if you remove the responsibility for the correctness and timeliness of the input information from the gatherer and user, you get a much higher error rate. This is the problem mentioned in an earlier chapter with central data control and collection groups. Thus, standards must exist that make the transfer of information from one system to another feasible.

The answer that can be given for the general case, without knowing the specifics of an organization is, yes, it would work with some qualifiers. Organizations using on-line data entry techniques, that allow for verification of information at the point of gathering, could establish procedures that would ensure the timeliness and accuracy of the information gathered. Thus, the responsibility for timely accurate information can remain where it belongs with the data gatherers. Thus, the organization must be in the stage of development where there are organizational standards for system design, programming, documentation and procedures. If the majority of line management (third level coordinators) is to be obtained from current line management personnel, then it is important that these people have a broader knowledge of the organization than would be reflected by their current duties. This might mean there is a need in additional training. The theory does assume that decision power does rest with the coordinator of a system, and subsystems must themselves be systems. Thus, it will not work in an organization where all decisions for all things is vested in the hands of one person, and everyone else has just the responsibility for managing the function. Managers would have to control the resources needed for their function. Finally, the number of changes needed is so great that the only way to accomplish the final structure without destroying the organization would probably be by phasing in various changes over a period of time. This would necessitate the modification of missions and goals until changes could be made.

The organizational structure proposed is not unique; other structures could be proposed that are not that radical a change from current structures. It would be informative to look at the same problem and modify the goals slightly and propose another structure.

In the discussion of the above structure the need for standards for information transfer between subsystems was emphasized several times. Let us assume the organization has not evolved all of these standards but does need planning information. The structure proposed previously would not work under these assumptions and the goals of an automated management information system, or the coordination of all resource management would be impractical. However, one can still obtain planning information and by using terminal input and having a college-wide master schedule for information, maintain a semiautomated management information system. The structure that would evolve is a structure that is currently found in some organizations. It would have four subsystems headed by vice presidents.

Vice President for Planning & Information Systems
    Planning & Budget
    Administrative System Design & Development
    Record Keeping & Reporting
        Accounting
        Registrar
        Purchasing
        Personnel
        Payroll

Vice President Marketing
    Admissions
    Development
    Sponsored Research

Vice President Administrative Services
    Plant Management
    Security
    Computer Center Operations
    Food Services
    Housing

Vice President Academic Services
    Learning Resource Management
    Instruction
    Research
    Counselling

If we examine this structure we can see that there are no uncoordinated processes and the only needed standard for information transfer is one that assures that no subsystem will be required to obtain new input information during its cycle. This can be accomplished by setting a college-wide cycle of data input for Record Keeping and Reporting, Admissions, Development, Sponsored Research, Plant Management, Food Services and Housing. The actual cycle time for information update can be as short as the administration feels is needed. If all offices of the college are equipped with terminals, this period could be one day.

With this organizational structure a college-wide calendar which lists all events is crucial for proper coordination between subsystems. In order to obtain the objective of being a multi-mission institution, the cycle times established for each subsystem would have to be reflected in this calendar. The calendar becomes a global item for each of the four subsystems. The period for a cyclical update of the calendar should be established with the objective of obtaining as much flexibility as is needed to meet the multiple missions of the college.

Obviously this proposed structure is not the only possibility. Whatever structure the organization eventually evolves may be less important than the ability to analyze what they feel must be coordinated and the setting of practical cycle times that will allow everyone to do their jobs. Neither suppliers nor customers have been emphasized in the discussion above but they do play a large part in the system. Both proposed structures do take into account their role and the necessity to coordinate their interaction in the system. It might be an interesting exercise for the reader to apply the theorems and obtain an analysis of what rules for global items and what cycle times would have to be set to assure proper coordination in the system.

The concluding chapter of this book discusses how one would use this analysis technique to simulate a complex system.

VI
SOLUTIONS: MODELING A COMPLEX SYSTEM

The biggest advantage to this type of analysis when it is applied to an organization is that it can give the managers insights into where and why there are problems in meeting goals. Let's try the same technique to build a simulation model that allows us to evaluate an entirely new approach to a current problem.

SIMULATIONS

In order to have a useful model of a complex system, you must not only include all its parts, but you must also make sure you are looking at the correct system. As mentioned previously, in our current air transportation systems' airports are only a process module of the current system. Simulations have been used prior to the design of an airport, but they have turned out to be limited predictors. The reason is that you can model a particular airport as part of an airport maintenance system, and you can obtain information on how to meet certain objectives and goals: ease of baggage handling, ease of transfer between gates, accessible parking facilities. All these are desirable attributes. However, if you cannot control the volume of flights, or coordinate the placement of airport terminals and access to terminals over a fairly large land area, you often end up with a facility that is obsolete the day it is completed.

This is exactly what has happened with many of the world's air terminals. Dulles Airport, outside of Washington, D. C., was under-utilized for years, while National Airport is dangerous because of the volume of flights. The same situation exists in the New York Metropolitan area where Newark Airport is under-utilized and Kennedy Airport is over-utilized. The energy consumed in a month by stacked planes waiting to get down at the country's over-utilized airports could probably be used to light an entire medium-sized city for that month.

Needless to say, these airports are also dangerous and inconvenient to passengers. Near misses make them dangerous, and a passenger wanting to fly from New York to Washington, D. C., at the end of a business day can easily find himself needing four to five hours of travel time for a flight that takes less than an hour.

The other side of this coin is that there are sections of the country where little or no public transportation exists at all. There has never been the equivalent of a 'whistle stop' in our air transportation system. Large airports have always been built only around or near population centers. These airports have been used not only to serve the local populations, but also to serve as transfer points from small local airports. Because of the large airports' multiple missions, you get a mix of aircraft that is very difficult to control, and in rural areas you get very poor, unreliable service.

As in the earlier organization study, the system is complex and all the changes that might be needed once the system is fully understood might only be practical over a long period of time. However, again, it is worth looking at simply to get some insights.

A SYSTEM ANALYSIS TO CREATE A MODEL

Let's take the same approach that we did for an organization. In this case, we will do the preliminary system analysis for a hypothetical country where we can set objectives and goals and establish a model to test hypothesized solutions for meeting the objectives and goals. The coordination proposed for the model will follow the guidelines set by the theorems developed in earlier chapters.

We will compare the coordination structure developed for the model with the current structure in the U. S. for governance and regulation of the air transportation system.

Before we begin, let's review something about constructing these models that was mentioned in the summary examples of chapter IV. Assume for a moment that air transportation can be considered a system as defined in this book. There exists components within this system, mainly environmental factors, which are continuously interacting with the other entities of the system. However, it is only when the unusual happens that these factors will affect the system in ways that

SOLUTIONS: MODELING A COMPLEX SYSTEM 65

planning cannot control. For example, if you know an airport normally receives 24 inches of snow in January, you can plan to have proper snow removal equipment. However, if it rarely snows and, one February, as happened February 1979 in Washington, D. C., you receive 24 inches in one day when normally the accumulation for the month is less than six inches, the event will definitely affect the interactions in the system.

Using the same arguments as in the examples in chapter IV, we can assume that the environment can be handled as a subsystem and air transportation becomes a good candidate for a discrete event simulation.

The environmental factors that interact with the system can be treated as a subsystem in the model of the system in spite of the fact that they obviously continuously interact with the system. Their overall influence on the system will be testable by controlling the number and type of special events that they generate.

This will be the type of model where the goals and the objectives determine the structure of the model. There will be many assumptions made about the interactions of the various subsystems and the goal of the model will be to test these assumptions. Therefore, the first task in the analysis will be to set the goals and objectives for air transportation. These will be:

    1.   A safe system (i.e., minimize the probability of crashes).

    2.   A system that responsively serves all the people of the country regardless of their location.

    3.   A system which minimizes fuel consumption and, hence, is energy conservative.

    4.   A system which minimizes cost to the consumer.

    5.   A system which encourages the citizens to use the air transportation system only when that system is the most effective way of travel (i. e., minimizes cost, energy consumption, and time for the consumer).

What we want to show is how by using the system design guidelines that are set by the theorems, a model capable of coordinating the information needs and regulations of the system can be developed. The objective of the model would be to test various alternates for the placement of air terminals.

Let's assume for a moment that the airport facilities are process modules of the air transportation system, that their interaction is coordinated by a coordinator module, and that the totality is a subsystem of the system. What information would the transportation system's airport coordinator module need as entry INFOs from other parts of the system to transfer to its process modules? Let's do this task by identifying what information is needed for entry INFO and what should be set by the airport coordinator and its process module, keeping in mind that at the moment we are not talking about either the air controllers or the weather stations, but just the facility itself.

AIRPORT FACILITIES

We have assumed that the airport facilities are process modules that are to be coordinated by one coordinator module. This of course is not the current situation in the U. S. Agencies of each state control airport placement and goals. Let's examine this more closely. If the airport facilities simply accommodate air traffic, passengers, suppliers and visitors by some predetermined rules set by the coordinator module, is this practical? To answer this question, simply remove the external interactions (that is, the passengers, the air traffic, suppliers and visitors) and make your objective that they be ready in case these should appear some day.

Runways must be maintained even if no one lands on them because weather will have its effect on the surface. Similarly, buildings must be maintained. Personnel has to be hired and trained. Preventive maintenance schedules have to be determined. Thus, there is a cycle time for this operation even if the facility is not used, and the cycle is shorter than the planning cycle needed by a coordinator of all facilities. Theorem 8 tells us that in this situation you could handle each airport as a subsystem coordinator. Let's explore this further and see what entry INFO would be needed by both the coordinator of all airport facilities and the coordinators of each facility.

COORDINATOR OF ALL AIRPORT FACILITIES' RESPONSIBILITIES, OBJECTIVES AND INFORMATION NEEDS

If the structure we are going to consider is one where each facility is a subsystem, then the cycle time of the coordinator module for all of the facilities must be greater than or equal to the cycle time of the airport subsystem (theorem 1). The subsystem's airport facilities will have to coordinate airport maintenance and day-to-day operation—this does have a shorter cycle time than the planning function for all airports. Therefore, we can establish the following responsibilities and objec-

SOLUTIONS: MODELING A COMPLEX SYSTEM

tives for the coordinator of all airport facilities:

This coordinator module should *not* be responsible for facility maintenance or day-to-day operations of each facility.

The coordinator should be responsible for guidelines and regulations that address the objectives of safety, security and responsiveness and energy conservation. These can be used by the coordinator for periodic audits of the facilities. Although the coordinator should *not* be responsible for planning for the operation of the facility, it should be active coordinating the planning of the location of new facilities, and phasing out of existing facilities. Keep in mind the objective of this planning for facilities should be cost effective, safe, energy conservative, responsive public transportation available to all residents of the country.

In order to ascertain what place in the hierarchical structure of our model this coordinator of all facilities should be, let's examine the information needs of the coordinator for the planning function outlined above.

INFORMATION FROM INTERNAL OPERATIONS OF EACH FACILITY

- Number and types of current air travelers.
- Number and types of flights.
- Types of aircraft needed (i.e., size, landing, and take-off needs).
- Fuel consumption of current flight patterns.
- Safety records of current flight patterns.
- Non-productive flying times in current flight patterns (stacking, redirecting).
- Factors causing redirecting flights and closing of facilities.
- Costs associated with current facilities.
- Suppliers associated with current facilities.
- Terminal's use information (baggage handling, customer flows, gate utilization, etc.)
- Terminal access by passengers, suppliers and visitors.

The control of much of this INFO would not be in the hands of the coordinator of airport facilities as we understand the function today. If the responsibility for planning is to be at this level, then let's examine what functions would have to be coordinated in order that plans might be implemented.

The coordination and/or monitoring of the following would be necessary:

- Airport facilities
- Air controllers
- Airlines using the facilities
- Air transportation-related communications
- Local weather patterns
- Local access to airport facilities
- Communications

What is being indicated, of course, even in this sketchy analysis of the situation is that simply to meet the information needs, coordination must occur both at the level that the airport module is placed, and above this level by coordinators who have the ability to view the entire transportation, energy, weather, and communication capability of the country.

Theorems 13 and 14 can now be used to guide us in a structure. Theorem 13 deals with the relationship between entry information to a subsystem and hierarchical placement. The airport facility subsystem will have significant events occur when something out of the ordinary occurs with all the other entities listed above.

Theorem 14 tells us that if a decision is to be made by entry INFOs, then the value of the switch or flag must be set hierarchically above the first level coordinator module of the subsystem making the decision or it must be determined externally to the system. If the coordinator module that controls entry information to the subsystem airport facilities also coordinates the other entities, then it could efficiently determine which entities it must access when a significant event occurs.

This coordinator module could also determine the constant value of global INFO (theorem 15) because its cycle time would be greater than the cycle time of all other entities it coordinated.

Changes to the attributes of INFOs could be made without too much danger (theorem 16) of forgetting to change global information if this design structure was adapted.

The current organizational structure of the U. S. Government is such that the information needed by a coordinator of air terminals would come from at least the following regulatory agencies or departments:

CAB
FAA
FCC

SOLUTIONS: MODELING A COMPLEX SYSTEM   69

DEPT OF COMMERCE
DEPT. OF TRANSPORTATION
DEPT. OF ENERGY
DEPT. OF THE INTERIOR
LOCAL AND STATE AIRPORT AUTHORITIES
LOCAL AND STATE DEPTS. OF TRANSPORTATION

To meet the system objective of cost effective, safe, energy conservative, responsible public transportation available to all residents of the U.S., the standards for reporting would have to exist and information of all these agencies would have to be coordinated. No one in the country, not even the President, has the authority to coordinate or set standards for all these agencies.

Fortunately we are dealing with a mythical country. So let's try to discover if air transportation can be modeled as a subsystem of something called the transportation system. In order to meet all the goals we outlined at the beginning of this chapter, all transportation facilites of the country would have to be looked at. Thus this is an important question to answer.

To answer this, we must look at the planning needs at the next level of hierarchy in a system, so far as we have proposed;

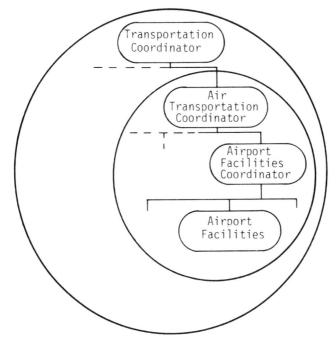

The next level of coordination would need information from:

air transportation
road transportation
rail transportation

Both road transportation and rail transportation need weather information and communication capability, as does air transportation. The road and rail transportation continuously interact with air transportation. Obviously if there is a breakdown in any of the systems, supplies and passengers for all systems are affected. Thus, one has to conclude that these are not proper subsystems of a system called the transportation system of the country.

However, a discrete event simulation model could be constructed where each of the modules rail transportation, road transportation, and air transportation are treated as subsystem of the transportation system. As discussed in chapter IV, the environmental and communication factors could also be handled as subsystems. Each of the subsystems would then create special events that the coordinator module of the entire system could use to record information from each of the subsystems to test interactions.

In fact, we have shown air transportation can be modeled separately as a system. If you took this approach, then it would have to be assumed that entry information from rail and road would be constant for the entire cycle of the system. This assumption would have to be made because of the definition of a system.

Often in proposing an organizational structure or analyzing a system, we get fooled into constructing structures that are unwieldy because of the names we have given items. We do not look at their functions or attributes, and based on the name itself we assume that the function is a system. What this preliminary analysis has pointed out is that in doing the analysis if you have some guidelines you can ignore terminology and simply look at how best to handle the problem. In this preliminary analysis for constructing a model to test interactions, we have already seen some indication why planning for many transportation needs in the country is difficult, if not impossible.

If we assume that better mass transportation is one of the answer to our energy problem, and that something really has to be done about solving this problem, then this preliminary analysis points out that one thing that definitely should be looked at is coordination of the planning for the location of the airport facility. The interstate highway system in the U.S. shows that coordination in planning does not necessarily mean a transfer of control from the states to the federal level. Planning would probably be more efficient if control resided on the federal level, assuming the federal government was truly a coordinated hierarchical system. It has been pointed out in the example for theorem 10 this is probably not the case.

It is important to point out at this time that some of the objectives and goals set for the system might be in conflict with the objectives and goals of other systems within this hypothetical country, or the goals and objectives of the country as whole. There is really no way of testing this unless the entire country was looked at as a system, and objectives and goals set for the entire country could be evaluated, and whether or not transportation is really a proper subsystem of the system 'The Country' could be determined. However, in modeling to gain insights, it is not necessary to want to get the perfect solution. You can and probably should model transportation and even air transportation as a system just to get some insights into the interactions that occur. The problem with using modeling techniques to obtain insights is that people then tend to forget all the approximations and guess they made, and use the models for political ends as if they were a true reflection of reality. Rather than proceeding with this analysis, as its complexity will lead us to areas of expertise way beyond the scope of this book, let's see what conclusions can be drawn from the examples presented in both this chapter and the preceding chapter.

CONCLUSIONS

The system analysis technique demonstrated in Chapters I-IV of this book can be applied to real problems. They are particularly helpful to isolate where coordination must occur in a system, and how and where information must be collected and transfered. However, there are qualifiers to the usefulness of the insights obtained using this type of analysis or, for that matter, any other type of structured or theoretical analysis technique.

Similar to the qualifiers that had to be placed on the system analysis of an organization, human judgement, helped by tools such as models and appropriate system analysis, is the only good guide to problem solutions. Any of the problems that we want to apply system analysis techniques to are so complex that the best analysis and/or model will only be an approximation. As mentioned in Chapter I, humans cannot be considered a subsystem using these definitions. Yet, in everything we have examined, they are a crucial and variable element in whether or not any of the proposed solutions will work. Thus, if you want to reorganize an organization, or model a complex system, you can use system analysis techniques to develop a proposal. However, it is a very foolish individual who would implement that proposal without making an effort to get the opinions of the people who must operate in the environment proposed. As it turns out from experience, even that is not enough. Valid reactions to a proposal often do not come till after the implementation. Thus, a system has to be designed so that it is modifiable without disrupting or destroying the entire system.

Disruptions caused by change can be very costly to an organization or to a naturally occurring system. Yet, where dislocations and negative impact will occur are often the hardest things to predict. Using the theorems, it is possible to analyze how coordination is occurring under current procedures. You'll want to trace information flow; see where under the current procedures the hierarchial structure coordination is occurring; and find out where there are proper subsystems and what units are currently acting as modules.

The next step is to analyze what the proposed changes will do to the current structure. If there are proper subsystems, and the proposed changes are so radical that they actually change the steady state of a subsystem or the entire system, it is important to estimate both the costs of the changes and the benefits. In an organization the type of costs that might occur are not just the obvious ones of new equipment or services, but costs associated with high absenteeism, turnover, and grievances. If the benefits outweigh the costs, then the impact of change can be mitigated by including the people who will be affected by change into the planning for the change.

The impact of changes that are occurring externally to the system can also be analytically analyzed. Changes in government regulations are an example of this type of external change that often impacts the system. Industry frequently argues that these changes are lowering productivity and raising cost, but no one identifies precisely where or how. If an organization has achieved a steady state and a new regulation forces major modifications in information flow, it will affect the structure, often making it difficult to comply with the regulation. Sometimes the changes to the organization caused by the new regulation will cause short term problems and result in long term benefits, but only analysis of the system as a whole will point out this potential.

The impact of change can be modified if the changes are planned over a period of time. An analysis of the organization which identifies information and process needs, establishes proper coordination and

SOLUTIONS: MODELING A COMPLEX SYSTEM

identifies subsystems and modules can be used to plan for new goals while minimizing impact.

Although no emphasis has been placed on structured analysis and design of automated systems in this book, there is a relationship between these theorems and the goals of that technique. In order to construct a well-structured automated system, you must trace information flow, identify cycle times of information and processes, find the requirements of processes and information for coordination, and identify proper hierarchical structure. The theorems can be used as guidelines on how to achieve these goals. They reinforce the advice that has been given on structured design and show how hierarchical structure and placement affects the overall behavior of the automated system. They provide the analyst with clues as to where it is appropriate to break the system into modules or subsystems.

However it should be pointed out that a well-structured automated system that does not mesh well with the structure of the organization will often be unused. This happens when the changes in the organization needed to effectively use the automated system require an entity in the organization to change from a proper subsystem to a coordinator module, or when coordination for maintenance of the information in the automated system does not exist in the organization. This is why so many management information systems have failed to be implemented.

A well-structured automated system is certainly desirable. It is easier to modify and to test, and thus less expensive for the organization to maintain. However, it is only a desirable characteristic for the organization system. It is not a necessary factor for successful automation. The reason is that the computer can be one of the most valuable resources an organization can use, but it is simply that, a resource tool for the employees, customers, and suppliers of the organizations. When that fact is forgotton and an attempt is made to impose automation with no coordination of information flow, the results are often frustrating.

To illustrate how the theorems are related to automation, let's look at some hypothetical goals of an organization, and the most basic design problem in automation, the design of the computer's hardware and software. Let's assume that an organization is planning to install a computer and hopes to achieve several goals among which are security of information and processing that contains a high percentage of calculations and comparisons. They have decided that to achieve a high level of security it is important to transfer the information temporarily stored on disk packs to tape after certain processes are completed.

A problem that computer system manufacturers have been trying to address since the first computer was built for commercial applications is how do you get reasonable utilization of the central processing unit, which operates in nanoseconds, when the information can only be input or output on much slower devices. The added requirements on input and output created by security requirements and terminal usage make an answer to this problem even more important.

What we want is to look at the overall system design for the computer's hardware and software. Obviously a second generation machine that reads in information, processes it and then prints it one task at a time is not the answer to a cost effective solution for the example's processing goals. Let's look at a more modern machine that queues up jobs. By imposing a fixed artificial cycle time, breaking the operating system into subsystems (one of which coordinates input/output (I/O)) and handles all coordination through the executive of the operating system on the main processor of the computer the designer attempts to optimize processor usage by handling other tasks during I/O interrupts.

The theorems tell us this is not the answer. You are dealing with a system whose cycle time has to be large enough to account for the slowest device on the system. The processor is the fastest device on the system and has the shortest cycle time. Yet, you have imposed a second level subsystem with a much longer cycle time (the I/O subsystem of the operating system) on this processor. To address this problem of slow I/O, an intermediate artificial cycle time is imposed to check on the status of the I/O processes.

In order to do this checking of task status, it becomes essential to store a great deal of information about the tasks on the main memory. The operating system continues to grow in complexity and size as more and more different I/O devices are added. You now have the situation where the operating system is never free of errors and the overhead of keeping track of what you are processing grows and grows. You can mitigate the situation slightly by adding an independent processor for handling terminals, removing some of the decision making from the main processor and hence some of the overhead.

Looking at cycle times, information needs, and the type of processing needed to meet the goals, it is possible to propose a better design. This is particularly true at this point in time, because the price of microprocessors is so low that you can even build some redundancy into a system when the tasks are subdivided into small manageable entities.

Let's assume a computer system is constructed in the following

SOLUTIONS: MODELING A COMPLEX SYSTEM

way. There is a microprocessor that serves as the coordinator module of the system. Its job is to decide when and what information must be transferred between the processor that handles arithmetic and comparison tasks of application programs (which we will refer to as the main processor and is a coordinator module of the system) and separate I/O processors which are handled as a subsystem with a fixed cycle time. The main processor when it recognizes an I/O instruction immediately transmits the information needed to codify and execute the instruction to the microprocessor that is serving as the system coordinator. The microprocessor then either sends back information on completed I/O tasks and their priority and/or initiates new tasks depending on its monitoring of current status tables of the main processor. The information on what to do next is transmitted when an acknowledgement of receipt of the I/O message is sent.

I/O is treated like a subsystem by the coordinator module of the system. The processors are queried on a cyclic basis for the status of their tasks. The actual cycle time of the I/O subsystem would be determined by the speed of the devices and the mix of jobs and priorities in a particular installation.

Obviously the analysis would have to go much deeper before an actual design could be achieved. All we have done is remove the executive of the operating system from the main processor to a coordinating processor and coupled two processors together allowing for a constant flow of information between the two. In addition we have adopted the time slicing device of interacting computers for use with an I/O subsystem. Since the objective of the example was only to show how the theorems affect the design of an automated system, we will stop here.

Having the potential efficiency of a uniquely tailored computer system does not necessarily mean that the organization using the system will benefit from the advantage. The analyst who constructs the application programs to be used on the computer system must take the requirements of the structure of the computer system into account in his or her design of the application system. Even that is not enough to insure success. Knowing how the design will impact the people responsible for maintaining the information is an important factor in proposing an appropriate design for the application system. Thus, a true top down approach does not start with the design of the application programs. It starts with an analysis of the entire organizational system and a study of information flow that takes into account the proper roles of all the people and the equipment that are members of the system.

Thus, this book ends on a cautionary note. System analysis is a powerful tool for gaining insights into the interactions of complex systems, but value judgements must be left to humans, and cumbersome, unethical, or amoral systems are the results of poor value judgements, not necessarily bad system analysis.

BIBLIOGRAPHY

American Management Association: "Establishing an Integrated Data-processing System, Special Report," New York, 1956

Argyris, Chris, "Understanding Organizational Behavior," The Dorsey Press, Inc., Homewood, Ill., 1960

Ashby, W. Ross, "An Introduction to Cybernetics," John Wiley and Sons, Inc., New York, 1956

Balentfy, N., and Burdick Chu, "Computer Simulation Techniques," John Wiley, New York, USA, 1966

Barish, Norman N., "System Analysis for Effective Administration," Funk & Wagnalls Company, New York, 1951

Boulding, Kenneth, "General System Theory: The Skeleton of Science," Management Science, April 1956, pp. 197-208.

Bonini, Charles P., "Simulation of Information and Decision Systems in the Firm," Prentice-Hall, Englewood Cliffs, 1963

Bright, James R., "Automation and Management," Harvard University, Graduate School of Business Administration, Boston, 1958

Bross, Irwin, D.J., "Design for Decision," The Macmillan Company, New York, 1953

Burlingame, John F., "Information Technology and Decentralization," Harvard Business Review, November-December, 1961, pp. 121-126

Chu Kong, and Naylor, Thomas H., "A Dynamic model of the Firm," Management Science, XI, May 1965, pp. 736-750

Churchman, C. West, "An Analysis of the Concept of Simulation," Symposium on Simulation Models, A.C. Haggatt and Frederick E. Balderson (eds.), South Western Publishing Co., Cincinnati, 1963

Drucker, Peter F., "Men, Ideas and Politics; and Management: Tasks, Responsibilities, Practices," Harper and Row, New York, 1974

Drucker, Peter F., "Management for Results," Harper and Row, New York, 1964

Drucker, Peter F., "The Practice of Management," Harper and Row, New York, 1954

Drucker, Peter F., "Managing for Business Effectiveness," Business Classics: Fifteen Key Concepts for Managerial Success, Harvard College, 1975, pp. 58-65

Eckman, D. (ed.), "Systems: Research and Design," John Wiley & Sons, Inc., New York, 1961

Gallager, James D., "Management Information Systems and the Computer," American Management Association, Inc., New York, 1961

Goldman, Stanford, "Information Theory," Prentice-Hall, Inc., Englewood Cliffs, New Jersey, 1953

Greenberger, Martin (ed.), "Management and the Computer of the Future," The MIT Press and John Wiley, Cambridge, Mass., New York, 1962

Guest, Robert H., "Quality of work-life-learning from Tarrytown," Harvard Business Review, July-August, 1979, pp. 76-87

Haire, Mason, "Psychology in Management," McGraw-Hill Book Company, Inc., New York, 1956

Herzberg, F., "One More Time: How do you Motivate Employees?" Business Classics: Fifteen Key Concepts for Managerial Success, Harvard College, 1975, pp. 13-22

Hopkins, Robert C., "Possible Applications of Information Theory to Management Control," IRE Transactions on Engineering Management, March 1961, pp. 40-48

Jaques, E., "Taking time seriously in evaluating jobs," Harvard Business Review, September-October 1979, pp. 124-132

Johnson, R.A., Fremont, K.E., Rosenzweig, J.E., "The Theory and Management of Systems," McGraw-Hill Book Company, Inc., New York, 1963

Kaimann, R.A., "Structured Information Files," John Wiley & Sons, Inc., New York, 1973

Kantrow, A.M., "Why read Peter Drucker," Harvard Business Review, January-February 1980, pp. 74-82

Laszlo, E., "Introduction to system philosophy," Harper & Row, New York, 1972

Lieberman, Irving J., "A Mathematical Model for Integrated Business Systems," Management Science, July 1956, pp. 327-336

Litterer, Joseph N., "The Simulation of Organizational Behavior," Journal of the Academy of Management, April, 1962, pp. 24-35

March, James G., and Simon, Herbert A., "Organization," John Wiley & Sons, Inc., New York, 1958

Martin, Elizabeth (ed.), "Top Management Decision Simulation: The

AMA Approach," American Management Association, Inc., New York, 1957
McMillan, Claude & Gonzalez, Richard, "System Analysis," Richard D. Irwin, Inc., Homewood, Ill., 1965
Meadow, Charles T., "Man Machine Communication," John Wiley & Sons, Inc., 1970
Medina, B., "Designing Automated Systems to ensure security and privacy," Information Privacy, Vol. 1, No. 5, May 1979, pp. 203-208
Medina, B., Picciano, A., "Committee Approach to User Involvement in Establishing of a College Information System," Cause, Innovating System? Solution or Illusion?, Vol. II, Proceedings of the 1974 CAUSE National Conference, Arlington, VA 1975, pp. 261-270
Medina, B., "User Involvement in Establishing a College Information System," CAUSE 1975, National Conference, Denver, Colorado, December 1975, pg. 698
Medina, B., "Designing Simulation Models for Ease in Testing and Validation of Interacting Systems," Proceedings Summer Computer Simulation Conference, Simulation Councils, Inc., La Jolla, California, 1973, pp. 213-215
Naylor, Balintfy, Burdick, Chu, "Computer Simulation Techniques," John Wiley, New York, USA, 1966
Orr, William D. (ed.), "Conversational Computers," John Wiley, New York, 1968
Parnas, D.L., "On the Criteria to be Used in Decomposing Systems into Modules," Carnegie-Mellon University, Communications of the ACM, December 1972, Vol. 15, No. 12
Simon, Hubert A. and Newell, A., "Heuristic Problem Solving, The Next Advance in Operational Research," Operations Research, January-February, 1958, pp. 1-10
Strauss, George and Sayles, Leonard: "Personnel: The Human Problems of Management,: Prentice-Hall, Inc., Englewood Cliffs, New Jersey, 1960.
Taubes, Mand, Wooster, H., "Information Storage and Retrieval," Columbia University Press, New York, 1958
Tannenbaum, R., Schmidt, W.H., "How to Choose a Leadership Pattern," Business Classics: Fifteen Key Concepts for Managerial Success, Harvard College, 1975, pp. 115-124
Tocher, K.D., "The Art of Simulation," D. Van Nostrand Co., Princeton, New Jerset, 1963

von Bertanffy, L., "General System Theory: A New Approach to the Unity of Science," Human Biology, December, 1951, pp. 303-361

Walton, Richard E., "Work innovations in the United States," Harvard Business Review, July-August 1979, pp. 88-98

Weiner, Norbert, "The Human Use of Human Beings," Houghton Mifflin Company, Boston, 1954

Weizenbaum, Joseph, "Computer Power and Human Reason, from Judgment to Calculation," W.H. Freeman and Company, San Francisco, 1976

INDEX

air transportation system
   used as an example 4, 18, 48, 63
attribute
   INFO attribute definition 3
attribute table 3
Boulding, K. x, 77
bounded 4
branch
   definition of 3, 20
closed 4
college's organization
   used as an example 32, 47-62
constant
   INFO attribute 3
coordination
   of functions 5
   of information 27
coordinator module
   definition 2
   number and level 20
   number in system 21, 22
   relation to subsystem 18
   relation to system 17
computer system's structure 74
computer center
   used as an example 25, 32
corollaries to
   theorem 5, 19
   theorem 15, 39
cycle time
   definition of 2
   subsystem 9
   system 9
decision table
   INFO attribute 3
decomposition
   subsystem 16
   system 6
entity
   definition of 2
entry
   importance of 2
   system key word 8
exit
   importance of 7
   system key word 8

file
   definition of 3, 38
flag
   definition of 3
   placement of 38
forest
   used as an example 7, 44, 46
global INFO
   definition of 3, 38
hierarchy
   definition of 2, 10
   subsystem 10
   used in example 13
INFO
   calculation of 19
   coordination of 32
   definition of 3, 17
   described 36
   hierarchical subsystem relation 37
information (see INFO)
interaction
   importance 7
   system key word 8
Jaques, E 55, 78
Johnson, R IV, x, 78
Laszlo, E VI, x, 78
level
   definition of 2
logical relationship
   INFO attribute 3
management information systems
   used as an example 36
module
   definition of 2, 16
NCHEMS viii
organizations as systems
   used as an example 5, 13, 19, 25, 37, 39, 47-62
probablistic relationship
   INFO attribute 3
process modules
   definition of 3, 17
   relation to subsystems 18
proper subsystems
   definition of 9
relational relationships

INFO attribute 3
Rosenzweig, J x, 80
schematics
   coordinator module 19
   process module 19
   system 12
simulation
   discrete event 42
   used as example 42, 63-76
single entry
   importance 7
   system key word 8
single exit
   importance 7
   system key word 8
steady state
   definition of 3, 37
   description of 37
   used in an example 31, 45
submarine encounter model
   used in an example 43, 45
subsystem
   cycle time of 2
   definiton of 2
   hierarchical order of 10
   hierarchical relation to INFO 37
   inclusion within system 10
   relation to coordinator module 17
   relation to number of coordinator modules 21
switch
   definition of 3
   placement of 39
system
   definition of 2
   one coordinator module 21
theorem 1
   example of application 30, 55
   proof of 9
   statement of 9, 29
   used in proof of theorem 2, 10
   used in proof of theorem 13, 37
theorem 2
   example of application 31
   proof of 10
   statement of 10, 28
   used in proof of theorem 12, 24
   used in proof of theorem 13, 37
theorem 3
   example of application 31, 55
   proof of 17
   statement of 17, 28
   used in proof of theorem 4, 18
   used in proof of theorem 7, 21
   used in proof of theorem 11, 23
theorem 4
   example of application 31
   proof of 18
   statement of 18, 29
   used in proof of theorem 7, 21
theorem 5
   corollary 19
   example of application 32
   proof of 19
   statement of 19, 29
   used in proof of theorem 6, 20
   used in proof of theorem 7, 21
theorem 6
   example of application 33
   proof of 20
   statement of 20, 29
   used in proof of theorem 7, 22
   used in proof of theorem 8, 22
theorem 7
   example of application 34, 53, 55
   proof of 21
   statement of 21, 29
   used in proof of theorem 8, 22
   used in proof of theorem 9, 22
theorem 8
   example of application 34
   proof of 22
   statement of 22, 29
theorem 9
   example of application 34
   proof of 22
   statement of 22, 29
theorem 10
   example of application 34
   proof of 23
   statement of 23, 29
theorem 11
   example of application 34, 55
   proof of 23
   statement of 23, 29
   used in proof of theorem 12, 34
theorem 12
   example of application 34
   proof of 24
   statement of 24, 29
theorem 13
   example of application 37, 68
   proof of 37
   statement of 37, 41
   used in proof of theorem 14, 38
theorem 14
   example of application 38, 68
   proof of 38
   statement of 38, 41
theorem 15
   corollary 39
   example of application 39, 68
   proof of 39
   statement of 39, 41
theorem 16
   example of application 45
   proof of 40
   statement of 40, 41
von Bertanffy x, 80
weather systems
   used as an example 14
Weizenbaum x, 80